T0200372

High-Speed 3D Imaging with Digital Fringe Projection Techniques

Optical Sciences and Applications of Light

Series Editor
James C. Wyant
University of Arizona

*Please visit our website **www.crcpress.com** for a full list of titles*

High-Speed 3D Imaging with Digital Fringe Projection Techniques

Song Zhang

Purdue University, West Lafayette, Indiana, USA

CRC Press
Taylor & Francis Group
Boca Raton London New York

CRC Press is an imprint of the
Taylor & Francis Group, an **informa** business

CRC Press
Taylor & Francis Group
6000 Broken Sound Parkway NW, Suite 300
Boca Raton, FL 33487-2742

First issued in paperback 2019

ISBN-13: 978-1-4822-3433-6 (hbk)
ISBN-13: 978-0-367-86972-4 (pbk)

Library of Congress Cataloging-in-Publication Data

Zhang, Song (Optical engineer), author.
 High-speed 3D imaging with digital fringe projection techniques / author, Song Zhang.
    pages cm -- (Optical sciences and applications of light)
 Includes bibliographical references and index.
 ISBN 978-1-4822-3433-6 (alk. paper)
    1. Three -dimensional imaging--Technique. 2. Image processing--Digital techniques.
3. Diffraction patterns. I. Title. II. Title: High-speed three-d imaging with digital fringer projection techniques.

TA1560.Z45 2015
006.6'93--dc23   2015034635

Contents

Series Preface

O PTICS AND PHOTONICS ARE enabling technologies in many fields of science and engineering. The purpose of the Optical Sciences and Applications of Light series is to present the state of the art of the basic science of optics, applied optics, and optical engineering, as well as the applications of optics and photonics in a variety of fields, including health care and life sciences, lighting, energy, manufacturing, information technology, telecommunications, sensors, metrology, defense, and education. This new and exciting material will be presented at a level that makes it useful to the practicing scientist and engineer working in a variety of fields.

The books in this series cover topics that are a part of the rapid expansion of optics and photonics in various fields all over the world. The technologies discussed impact numerous real-world applications, including new displays in smartphones, computers, and televisions; new imaging systems in cameras; biomedical imaging for disease diagnosis and treatment; and adaptive optics for space systems for defense and scientific exploration. Other applications include optical technology for providing clean and renewable energy, optical sensors for more accurate weather prediction, solutions for more cost-effective manufacturing, and ultra-high-capacity optical fiber communication technologies that will enable the future growth of the Internet. The complete list of areas where optics and photonics are involved in is very long and keeps expanding.

Preface

I N AUGUST 2013, I was approached by Ashley Gasque who asked whether I was interested in writing a book on 3D sensing technologies for the book series, *Optical Sciences and Applications of Light*, launched by Professor James Wyant, my academic hero. It was a great honor to be invited by Professor Wyant on this project; and the timing could not be better because I had just turned in my tenure package to Iowa State University, and was planning my career for the next stage. Writing a book about my decade of research experiences would be a great way to reflect what I have done in this field for over 15 years, and could provide some guidance to those who have not had such a long research experience.

The project turned out to be more difficult than I initially thought, mainly because I decided to move to Purdue University in December of 2014. Moving the whole family was not easy; and moving the family and the research laboratory together was a lot more difficult. As a result, this book project was substantially delayed.

The whole project was much more involved than I anticipated because the more I thought about the methods we developed to solve some problems, the more I felt that there could be more elegant, more rigorous, or more generic solutions to those problems. Therefore, this book was developed along with exploring and experimenting with new technologies or methods within my research team. Furthermore, newer technologies also arrived, and thus new devices were purchased and experimented with to validate some concepts/thoughts discussed in the book.

With the advancements of modern computers, optically measuring 3D object surface topology becomes increasingly easier and faster. Due to the surface noncontact and non-invasive nature, optical methods have impacted numerous areas, including manufacturing, medical sciences, computer sciences, homeland security, and entertainment. The recent release of Microsoft Kinect® has enabled 3D technology to penetrate into our daily lives; and the combination of 3D additive manufacturing with 3D imaging technologies could generate additional excitement across disciplines.

As its title suggests, the book mainly focuses on digital fringe projection (DFP) technologies. However, it reviews various 3D imaging techniques in Chapter 1 and suggests that DFP techniques have some overwhelmingly advantageous properties comparing to other methods for many application areas; and thus concentrating this book on this specific topic would be of interest to many.

Chapter 2 lays down theoretical foundations of fringe pattern analysis techniques, mostly originated from laser interferometry, a still extensively adopted method in optical instruments. Generating digital fringe with digital video projection devices is then introduced in Chapter 3. This chapter elucidates the differences between three main digital video projection techniques: digital light processing (DLP), liquid crystal display (LCD), and liquid crystal on silicon (LCoS). Our experiments reveal that the choice of digital video projection technologies ultimately depends on final goals.

To help one master DFP techniques, the book covers a variety of core technical aspects ranging from how to properly unwrap phase maps temporally (Chapter 4) or spatially (Chapter 5); how to correctly generate fringe patterns with video projectors, knowing their nonlinear response (Chapter 6); and how to convert phase to coordinates through system calibrations (Chapter 7).

As a professor working in university settings, I found that the most effective way of teaching is to give hands-on examples. With this in mind, Chapter 8 provides a detailed example of building a 3D imaging system from scratch: from hardware selection to system design, testing, and calibrations. The book concludes in Chapter 9 with the pathway toward high-speed 3D optical imaging using DFP technologies. We acknowledge that even with tremendous advancements in the field, there are still numerous challenges that we ought to conquer to make such technologies more powerful and impactful.

There are many people to thank for making this book a reality. My first thanks go to Ashley and Professor Wyant, without their initial contacts, this book would not have been possible. I thank many of my former and current students who actually carried out the research including Tyler Bell, Chen Gong, Nik Karpinsky, Laura Ekstrand, Molly Lin, Beiwen Li, William Lohry, Yajun Wang, and Ying Xu. I extend my special thanks to Zexuan Zhu, a Purdue undergraduate, who has done a tremendous amount of work for this book. Chapter 8 was a result of his summer intern research in my laboratory. Without his help, the whole book project could have been further delayed.

I also thank my wife, Maggie Hao, for her continuous encouragement and support. Every time I felt I could not do this book, she was always the one to convince me that I could do it.

My son, Simon, was born in December 2012. I always felt that I had these two "*babies*" grow together over the past two years; and it is a true joy to be part of their growth.

This book summarizes insights that I have gained, and the lessons I have learned over the past 15 years. I hope that this book will be a reference for engineers and scientists to learn and master 3D imaging with DFP techniques, and to substantially shorten their learning curves. To further facilitate their learning, some source codes are provided for the users to play with.

Author

Song Zhang is an associate professor of mechanical engineering at Purdue University. He earned his BS from the University of Science and Technology of China in 2000, his MS and PhD in mechanical engineering from Stony Brook University in 2003 and 2005, respectively. He then spent three years in the Mathematics Department at Harvard University before joining Iowa State University as an assistant professor of mechanical engineering. He moved to Purdue University in January 2015 as an associate professor of mechanical engineering.

Professor Zhang started working on 3D optical imaging in 2000 when he was a graduate student at Stony Brook University. He was often accredited with developing the first-ever high-resolution, real-time 3D optical imaging system. Professor Zhang has published over 130 research articles, edited one book, and contributed to a few other books. Many of his journal articles were featured on the covers and several were highlighted by the corresponding societies. Two of his papers were among the most cited papers during five-year periods for all papers published by those respective journals. Besides being extensively utilized in academia, the technologies he developed have been used by the rock band Radiohead to create a music video *House of Cards*; and by the Zaftig Films to produce the upcoming movie *Focus (II)*.

Professor Zhang has won a number of awards including the AIAA Best Paper Award, the Best of SIGGRAPH by Walt Disney, the NSF CAREER award, the College of Engineering's Early Career Engineering Faculty Research Award from Iowa State University, and the Forty under 40 Alumni Award from Stony Brook University. Due to his significant contributions to high-speed, high-resolution 3D optical sensing and optical information processing, he was elected as a fellow of SPIE—The International Society for Optics and Photonics.

CHAPTER 1

Introduction

THREE-DIMENSIONAL (3D) OPTICAL SURFACE topology measurement techniques refer to the process of digitizing physical object's surface topographical information using optical sensors. They are often called 3D optical metrology, 3D shape measurement, 3D optical sensing, 3D imaging, or 3D scanning; and different names are used by different communities, but they essentially refer to the same process.

With the advancements of modern computers, optically measuring 3D object surface topology becomes increasingly easier and faster. Due to the surface noncontact and noninvasive nature, optical methods have impacted numerous areas including manufacturing, medical science, computer science, homeland security, and entertainment. The recent release of Microsoft Kinect has enabled 3D technology to penetrate into our daily lives; and the combination of 3D printing technologies with 3D imaging technologies could generate additional excitement across disciplines.

3D imaging techniques have been improving over the past decades, and numerous methods have been developed including time of flight (TOF), stereovision, structured light, and digital fringe projection (DFP). Each of these technologies has been developed to conquer some limitations of some others and gained success for special applications. A recently edited book by Zhang [1] assembled the majority of optical 3D optical imaging methods, providing engineers and scientists a single resource to refer to. Instead of thoroughly presenting each individual technique, this chapter briefly overviews some of the most extensively adopted 3D imaging technologies, with a focus on the structured light technologies, which is closely related to the main subject of this book, DFP technologies.

1.1 TIME OF FLIGHT

Time-of-flight (TOF) 3D imaging technology is an optical method that is analogy to a bat's ultrasonic system [2,3] that timing the round trip of the light point leaving the emitter and bouncing back from the point surface to the sensor. Since the light travels at an approximately constant speed, the depth information can be determined from the time. However, the light travels at extremely fast speed ($\sim 3.0 \times 10^8$ m/s). This indicates that in order to resolve 1.00 mm depth change, the required timing sensor resolution is approximately 3.3 ps (10^{-9} s), making it difficult to be accurately achieved.

Instead of directly measuring the time difference, the TOF technology typically measures phase changes for the round trip by modulating the emitting light with a sinusoidally varying periodical signal, which can be mathematically described as [4]

$$S(i) = A(i)e^{j\phi(i)}, \tag{1.1}$$

where $A(i)$ is the amplitude of the signal and $\phi(i)$ is the phase information. Typically, the modulation frequency f_m is 20 MHz, and the depth z, to be determined, is proportional to the phase

$$z(i) = \frac{c}{4\pi f_m}\phi(i). \tag{1.2}$$

Here, c is the speed of light. To measure the encoded signal described in Equation 1.1, the TOF sensor shall simultaneously measure both the intensity and the phase information. Due to the complexity of manufacturing such sensors, even with great recent advancements in the past decades, the resolution of TOF sensors is still quite low ($\sim 320 \times 240$), and its depth resolution is also not high (mm). Even with such limited spatial and depth resolutions, because of their compact design, the TOF 3D imaging technologies still find their applications in automotive industry, human–computer interaction (such as the depth sensor for Microsoft Kinect 2), and robotics.

1.2 LASER TRIANGULATION

High-spatial and depth resolutions could be achieved by laser triangulation-based 3D imaging technologies. A laser triangulation system typically includes a laser light source, a detector, and a lens to focus the laser beam onto the detector. To perform a measurement, the laser emits a light onto the surface that reflects the laser beam to be sensed by the detector. If the emitting point of the laser source and the sensing point on the detector form a triangle, the depth information can be determined assuming that the system is calibrated.

To recover 3D information, a laser triangulation system usually requires the following major procedures: (1) calibrate the whole system; (2) segment the laser point or line centre from the background, which can be done through thresholding and/or background image subtraction; (3) apply the triangulation equation with the calibrated system parameters to calculate the (x, y, z) coordinates. Since laser triangulation systems could achieve high measurement accuracy and uses the single wavelength laser light source, they can be used to measure large objects (e.g., building, ships) for both out-door and in-door applications. Because of these advantageous features, laser triangulation techniques have been extensively studied and employed [5–7]. However, since it typically requires sweep a laser line or laser point to measure the whole surface, its measurement speed is quite slow, making it difficult to measure dynamically deformable objects. Moreover, since laser light is coherent, the speckle noise could influence the measurement accuracy.

1.3 STEREO VISION

Instead of point or line sweeping, the stereo-vision [8] technique uses two cameras to capture two 2D images from different viewing perspectives, and recovers 3D shape using triangulation that is similar to the laser triangulation method. The stereo-vision technique allows measuring the whole area at once without scanning, and thus is suitable for high-speed applications. The basic theory behind the stereo-vision technique is epipolar geometry and geometric optics, as illustrated in Figure 1.1. Instead of putting the lens in front of the camera sensor, as a real camera system does, the modeling typically put the camera behind the camera sensor to simplify mathematical descriptions, which is explained in detail in Chapter 7. A point (P) on one camera could be corresponding to many points on the other camera (e.g., Q^1, Q^2, Q^3, etc.) depending upon how far away a object point (e.g., Z^1, Z^2, Z^3) is since the projection from a 3D space (x, y, z) to a 2D space (u, v) is unique, but the inverse is a one-to-many mapping.

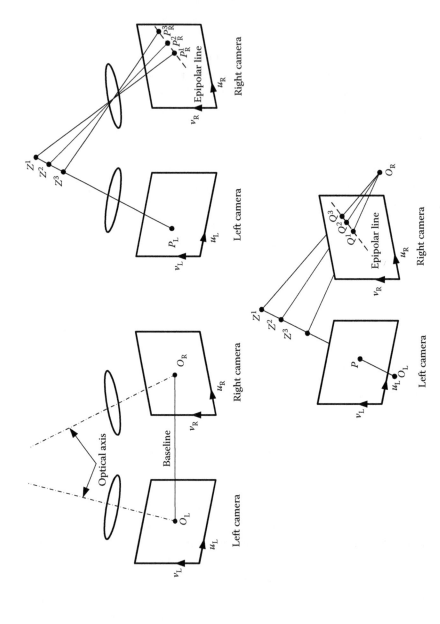

FIGURE 1.1 Epipolar geometry of a stereo-vision system.

The projection from a 3D space coordinate to a 2D image plane is typically represented as

$$
s \begin{Bmatrix} u \\ v \\ 1 \end{Bmatrix} = A[R, t] \begin{Bmatrix} x \\ y \\ z \\ 1 \end{Bmatrix},
\tag{1.3}
$$

where, s is a scaling factor that indicates depth of the 3D projection, R is a 3×3 rotation matrix, and t is a 3×1 translation vector. $[R, t]$ represents extrinsic properties of the system, the transformation from 3D world coordinate system to 3D lens coordinate systems, and

$$
A = \begin{bmatrix} f_u & \alpha & u_0 \\ 0 & f_v & v_0 \\ 0 & 0 & 1 \end{bmatrix}
\tag{1.4}
$$

represents the intrinsic properties of the camera including the focal length (f_u and f_v in u and (v) directions, respectively), α indicates perpendicularity of the camera horizontal and vertical pixels (typically 0 for modern cameras), and the principle point (u_0, v_0) representing the point that the optical axis intersects with the camera.

If both cameras are calibrated in the same world coordinate system, the stereo-camera system provides two sets of projection equations. The key to the success of a stereo-vision system is to find the corresponding point pairs from both cameras, once the corresponding pairs are established, (x, y, z) coordinates can be uniquely solved by using a least-squared method since there are six equations and five unknowns. In a stereo system, two images captured from different perspectives are used to detect corresponding points in a scene to obtain 3D geometry [8,9]. Detecting corresponding points between two stereo images is a well-studied problem in stereo vision. Since a corresponding point pair must lie on an epipolar line, the captured images are often rectified so that the epipolar lines run across the row [10]. This allows a method of finding corresponding points using a "sliding window" approach, which defines the similarity of a match using cost, correlation, or probability. The difference between the horizontal position of a point in the left image and that in the right image is called the disparity. This disparity can be directly converted into 3D coordinates.

Standard cost-based matching approaches rely on the texture difference between a source point in one image with a target point in the other [11].

The cost represents the difference in intensity between the two windows on the epipolar line and is used to weigh various matches. In a winner-takes-all approach, the disparity will be determined from the point in the right image that has the least cost with that of the source point in the left.

In addition to local methods, a number of global and semiglobal methods have been suggested [12–15]. One method that worked especially well was the probabilistic model named Efficient LArge-scale Stereo (ELAS) [16]. In this method, a number of support points from both images are chosen based on their response to a 3×3 Sobel filter. Groups of points are compared between images, and a Bayesian model determines their likelihood of matching. Since the ELAS method is piecewise continuous, it works particularly well for objects with little texture variation.

Stereo-vision methods, despite recent advances, still suffer from the fundamental limitation of the method: finding corresponding pairs between two natural images. This requirement hinders the ability of this method to accurately and densely reconstruct many real-world objects such as uniform white surfaces.

1.4 STRUCTURED LIGHT

The structured light technique has been extensively studied and used in the field of computer vision, machine vision, and optical metrology. Structured light is similar to the stereo vision-based method mentioned above except that a projector is used in place of one of the cameras [17]. The projector projects certain types of coded structured patterns to assist the correspondence establishment. For a structured light system, the structured pattern design is the first and one of the key factors to determine the ultimate achievable resolution, speed, and accuracy. This section summarizes a few extensively used structured pattern codifications.

1.4.1 Random or Pseudorandom Codifications

To establish the one-to-one mapping between a camera's pixel and the projector's pixel, one natural approach is to encode the projected image in a way such that the pixels are unique across the entire image in both the x and y directions; that is to say that every pixel can be labeled by the information represented on it [17]. Methods have been developed using such techniques as generating pseudorandom patterns or by using natural speckle patterns generated by a laser source [18].

In one such method, the pseudorandom binary array approach, an $n_1 \times n_2$ array is encoded via a pseudorandom sequence to ensure that any kernel $k_1 \times k_2$ on any position over the array is unique. The pseudorandom sequence to encode the $n_1 \times n_2$ array could be derived via the primitive polynomial modulo n^2 method that could be mathematically described as [19]

$$2^n - 1 = 2^{k_1 k_2} - 1, \tag{1.5}$$

$$n_1 = 2^{k_1} - 1, \tag{1.6}$$

$$n_2 = 2^n - 1/n_1. \tag{1.7}$$

A good example of the popular consumer product today that has taken a pseudorandom codification approach in its execution of computer vision is the Microsoft Kinect. The Kinect uses an infrared projector, a beam of infrared laser light passes through a diffraction grating to generate a set of random dots [20]. The pseudorandom dotted pattern is then captured by the infrared cameras, and because the projected pattern is predefined, a 3D point can be reconstructed by triangulation using the captured patterns.

Overall, benefits of using pseudorandom codification approach are that it is easy to understand and easy to implement for 3D imaging. However, it is difficult for such techniques to achieve high-spatial resolution since it is limited by both the projector and the camera spatial resolution in both u and v directions.

Since the aforementioned unique point correlation method provides six equations, and only five unknowns for this problem, solving (x, y, z) involves the use of a least-squares method. The overdetermined problem indicates that it is not necessary to have unique correlations in both u and v directions to reconstruct 3D coordinates. In other words, determining (x, y, z) coordinates from a structured light system only requires one additional constraint equation besides the calibrated system constraint equations (to be discussed in detail in Chapter 7). Therefore, the structured patterns can vary in one direction but remain constant in the other. This then eliminates the spatial resolution limits of both directions and, as a result, such techniques are extensively adopted in computer vision and optical metrology.

1.4.2 Binary Structured Codifications

Figure 1.2 shows the schematic diagram of a 3D imaging system using a structured light technique that contains stripes varying only in either the u or v direction but not both. The system includes three major units: the

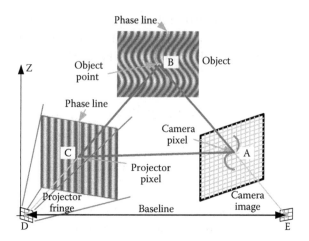

FIGURE 1.2 **(See color insert.)** Schematic diagram of a 3D imaging system using a structured light technique. (From S. Zhang, *Opt. Laser Eng.* 48(2), 149–158, 2010. With permission.)

image acquisition unit (A), the projection unit (C), and the 3D object to be measured (B). The projector shines vertical structured stripes straight onto the object's surface; the object's surface distorts these stripes from straight lines to curved ones if viewed from another angle. A camera then captures the distorted structured images from an angle that differs from the projection angle to form a triangle base. In such a system, correspondence is established through the structure codification information by analyzing the distortion of the known structure patterns. That is to say that the system knows which pattern is being projected and by determining and measuring the mutations of the pattern due to the object's surface, correspondence can be established.

Binary-coded structured patterns (only 0s and 1s) are extensively used in a structured light system as they are (1) simple: it is easy to implement since the encoding and decoding algorithms are straightforward; and (2) robust: it is robust to varying surface characteristics and noise because only two intensity levels (0 and 255) are used and expected [21]. In this method, every pixel of a pattern is encoded with 0 (e.g., 0 in grayscale) and 1 (e.g., 255 in grayscale). To achieve high spatial resolution, typically a sequence of binary structured patterns are projected and captured sequentially in time domain. Assuming the object is not changing over the period of time that all patterns are captured, the sequence of binarized patterns forms a sequence of 0 or 1 for each pixel, which defines a *code word* (a unique number that represents encoded correspondence information). Figure 1.3 illustrates

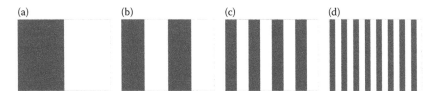

FIGURE 1.3 (a–d) A sequence of binary patterns for codification. To recover the code word, a simple comparison of whether the pixel is black or white is checked for each image, and the result is placed within the corresponding bit of the code word.

four binary structured patterns that can be used to uniquely determine eight code words. For the binary coded structured light method, the smallest differentiable pixel is the narrowest binary patterns used; and the narrowest stripes should be larger than one projector pixel and also larger than one camera pixel. The unique code word provides the unique correspondence between the projected image and the captured image along the direction perpendicular to stripes. Combining this constraint equation with projection equations leads to the solution of (x, y, z) coordinates for that particular pixel.

Spatial resolution for binary coded structured light methods is limited by both the projector resolution and camera resolution. Figure 1.4a shows a binary pattern, and Figure 1.4b shows its corresponding cross section. Here black represents binary level 0, while white represents binary level 1. For a stripe between point M and point N in Figure 1.4b, all points have the same intensity value, thus cannot be differentiated. Therefore, it cannot reach pixel level correspondence accuracy, making it difficult for this technology to reach very high measurement accuracy.

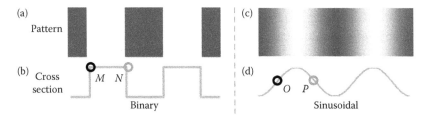

FIGURE 1.4 Comparison between binary and sinusoidal structured pattern. (a) Typical binary structured patterns. (b) Cross section of the binary pattern shown in (a). (c) Typical sinusoidal fringe pattern. (d) Cross section of the sinusoidal patterns as shown in (c).

The second drawback of the binary structured codification methods is that a large number of patterns are needed for codification since a binary code only uses two intensity levels, binary 0 and 1; this limits the maximum number of unique code words that can be generated to 2^n for n number of binary structured patterns. Therefore, to achieve dense 3D measurements, many binary structured patterns are required, making it less appealing for high-speed applications such as real-time 3D imaging.

1.4.3 N-ary Codifications

Although straightforward, robust to surface characteristics, and tolerant to noise, binary structured patterns have their drawbacks, especially when it comes to the large number of patterns needed for codification. To address this, without sacrificing spatial resolution, more intensity values could be utilized [22]. While introducing additional intensity values decreases a structured light codification's robustness, it increases its capturing speed; these trade-offs must be weighed for a specific application.

Instead of using only two intensity values (0 and 255) to create unique code words for every pixel in a pattern, the N-ary approach utilizes a subset of the range between these values; the most extreme case would be using all of the grayscale values (i.e., the full range 0–255). The code word for each pixel can be determined via the intensity ratio calculation [23,24]. The most basic form of this calculation assumes a spectrum of intensities, say 0–255, with linear "wedges" which may be placed along the vertical columns [17]. Two patterns are then projected onto the scene: one with the aforementioned wedge and one without (constant intensity). Next, an intensity ratio for each pixel can be derived with the values to calculate such a ratio being present in each of the two captured patterns. In alternative methods to this approach, many wedges are projected sequentially onto the scene while consistently increasing the wedge's period [24].

The intensity ratio method is good in terms of sensing speed as it needs fewer patterns for codification. However, this method, as well as the aforementioned methods, is more sensitive to noise (compared to the binary structured codifications), is still limited by the resolution of the projector, and is very sensitive to the defocusing of the camera and the projector. To overcome the limitations of the intensity ratio method, phase-shifting techniques can be introduced and various structured patterns are used such as the triangular shape [25] and the trapezoidal shape [26] patterns. These techniques can achieve camera pixel spatial resolution and are less sensitive to defocusing; however, they are still not completely immune to the defocusing problem.

It is important to note that for all these techniques, unlike the binary structure pattern methods, the nonlinearity of the projector needs to be considered in order to properly perform the measurement.

1.5 DIGITAL FRINGE PROJECTION

As noted by Zhang [21], the binary, the N-ary, the triangular, and the trapezoidal patterns eventually become sinusoidal if they are blurred properly. The blurring effect often occurs when a camera captures an image out of focus, and all of the sharp features of an object are blended together. Thus, sinusoidal patterns seem to be a natural choice. As shown in Figure 1.4c and d, the intensity varies across the image point by point. Therefore, it is feasible to reach pixel level resolution because the intensity values between horizontal neighboring pixels are differentiable. As the spatial resolution is high, the correspondence between the projector and the camera can be determined more precisely (camera's pixel resolution), this method allows for a higher accuracy.

Sinusoidal patterns (also known as *fringe patterns*) have the potential to reach pixel-level spatial resolution, and therefore have long been studied in optical metrology. The fringe patterns used in this instance are generated by laser interference. Instead of using laser interference where speckle noise could jeopardize the measurement quality, the DFP technique uses a digital video projector to project computer-generated sinusoidal patterns. In principle, the DFP technique is a special kind of triangulation-based structured light method where the structured patterns vary in intensity sinusoidally. Unlike the intensity-based method, DFP techniques use the phase information to establish correspondence and are typically quite robust to surface texture variations.

The phase could be obtained using Fourier transform profilometry (FTP) method [27] through Fourier analysis. FTP is extensively used for simple dynamic 3D imaging since only a single fringe pattern is required [28]. Such a method, however, is typically very sensitive to noise and surface texture variation. To improve its robustness, Qian [29] proposed the windowed fourier transform (WFT) method. Though successful, WFT still cannot achieve high-quality 3D imaging for general complex 3D structures due to the fundamental limitations of the FTP where the spatial Fourier transform is required. Modified FTP methods were developed to obtain better quality phase by using two patterns [30,31], but they are still restricted to measuring surfaces without strong texture and/or geometry variations.

To measure complex surfaces, at least three fringe patterns must be used [32]. If three or more sinusoidal structured patterns are used and their phase information is shifted, the pixel-by-pixel phase could be obtained without knowing neighboring information, thus making it immune to surface texture variations. These methods are typically referred to as phase-shifting methods.

The fringe projection technique originated from laser interferometry where the sinusoidal patterns are generated by the laser interference; and the majority theoretical foundation of fringe projection techniques came from laser interferometry. Therefore, the next chapter lays down the theories behind fringe projection; and the associated laser interferometry techniques is also briefly introduced.

1.6 BOOK ORGANIZATION

This book is organized as follows.

This chapter, or Chapter 1, introduces the motivation of this book, performed a brief literature review on optical 3D imaging methods, and presented the merits of the phase-shifting-based DFP techniques.

Chapter 2 lays down the theoretical foundation for phase-shifting methods. Phase-shifting algorithms originated from laser interferometry, and the theoretical foundation of physical optics for basic interferometry is introduced. The multiwavelength phase-shifting algorithms are also introduced.

Chapter 3 discusses the digital fringe generation techniques. DFP techniques take advantages of digital technology and can directly computer-generate sinusoidal patterns onto object surface. The key difference of a DFP system to a laser interferometer is the use of a digital video projector for fringe projection, and thus this chapter introduces the basics of three extensively used digital video projection methods, namely, digital light processing (DLP), liquid crystal display (LCD), and liquid crystal on silicon (LCoS) technologies. The differences among these technologies are explained and demonstrated experimentally.

Chapter 4 presents some special temporal phase-unwrapping methods that were developed for digital technologies, albeit the multiwavelength methods developed in laser interferometry can still be adapted for DFP systems.

Chapter 5 discusses general special phase-unwrapping methods with a focus on a rapid phase-unwrapping algorithm that we developed over the years.

Chapter 6 introduces the projector nonlinear gamma calibration methods. This chapter mainly focuses on the comparison of active versus passive nonlinear gamma calibration methods, and experimentally demonstrates that the active method should be adopted, if feasible, due to its robustness and immune to projector defocusing.

Chapter 7 discusses the importance of DFP system calibration. The fringe analysis methods only provide phase instead of coordinate information. This chapter covers the simple yet widely used reference-plane-based phase to coordinate conversion method. Alternatively, geometric calibration can be adopted to estimate system parameters for phase to coordinates conversion.

We believe that the DFP system geometric calibration is easier than ever before probably because the extensively adopted DFP system calibration method that we developed and available open source software packages (e.g., OpenCV) for parameter estimation. This chapter covers the basic principle of the calibration method we developed and generalizes the calibration approach to a system even when the projector is not in focus. The available open source software package makes it easy to adopt the newly developed calibration approach.

Chapter 8 presents an example of developing a DFP system from hardware system design, integration, calibration, and 3D shape recovery. The motivation of presenting such a hands-on example is to provide the readers the procedures to develop their own system through illustrations.

Chapter 9 illustrates the importance of a few key factors of DFP system design including the angle between the projector and camera, the fringe density, the fringe angle, and the fringe contrast. The main focus of this chapter is on one factor that is not generally considered, namely the fringe angle.

Chapter 10 overviews the state-of-art real-time and high-speed 3D imaging with DFP techniques and presents some challenges that still need to be conquered including accuracy improvement, 3D data compression, and 3D data postprocessing.

1.7 CONCLUDING REMARKS

This chapter only summarizes the 3D imaging technologies that are closely relevant to the DFP technique. Methods such as shape from focus or defocus [33,34], shape from texture or contour [35–38], digital hologra-phy [39–42], and coordinate measurement machine, are not discussed, and we encourage the readers to refer relevant literature on those technologies, if interested. Some of the materials presented in this chapter were modified from the previous publications [21,43,44].

CHAPTER 2

Theoretical Foundation of Fringe Analysis Techniques

F RINGE ANALYSIS METHODS and phase-shifting algorithms originated from laser interferometry, and the theoretical foundation for these methods are based on physical optics. This chapter introduces the principles of the fringe analysis methods and overviews some popular and extensively adopted fringe analysis methods.

2.1 INTRODUCTION

Fringe analysis methods for surface measurement starts with the *Newton's Rings* that is the interfering pattern between the testing surface and the reference surface and evolves to more advanced computerized fringe analysis methods for high-resolution measurements. Fourier transform [27] or WFT [29] methods were developed to retrieval-phase information from a single pattern to retrieve the phase through global or local Fourier transform. Fourier transform method and its variations [28] have seen tremendous success in measuring high-speed phenomena since it requires a single fringe pattern [45]. However, Fourier transform methods are typically limited to measure relatively "smooth" surface without strong texture variations. In contrast, phase-shifting-based techniques recover phase pixel by pixel, yet use a series of wave front-phase encoded fringe patterns.

The phase-shifting methods have been extensively used in high-accuracy and high-resolution 3D optical metrology because of the following major merits: (1) high-capture speed, since only three fringe images are required to recover one 3D frame; (2) high-spatial resolution, since the phase (and hence the depth measurement) can be performed pixel by pixel; and (3) less sensitivity to surface reflectivity variations since the phase calculation automatically cancels them out. Fourier or the phase-shifting methods typically only obtain phase values range from $-\pi$ to $+\pi$, and thus a spatial phase-unwrapping algorithm is typically required [46], which limits their capability of measuring discontinuous object surfaces or sharply changing object surfaces. Two- or multiple-wavelength phase-shifting algorithms were developed to increase its measurement capability. Overall, the applications of phase measurement have been extensively used in optical testing, real-time wave front sensing for active optics, distance measuring interferometry, surface contouring, microscopy, etc.

This chapter starts with the theory behind fringe pattern generation with laser interference (Section 2.2), continues to fringe analysis methods (Section 2.3), and concludes with the more powerful two- and multiwavelength phase-shifting techniques (Section 2.4).

2.2 FRINGE PATTERN GENERATION WITH COHERENT LASER INTERFERENCE

From physical optics, the wave front of a light source can be described as

$$w(x, y, t) = a(x, y)e^{i[\Phi(x,y)]}, \tag{2.1}$$

where x and y are spatial coordinates, $a(x, y)$ is the wave front amplitude, and

$$\Phi(x, y) = \frac{2\pi h(x, y)}{\lambda} \tag{2.2}$$

the wave front phase. Here λ is the wavelength, and $h(x, y)$ is the distance the light travels from its origin.

Assume there are two wave fronts, reference wave front, $w_r(x, y)$, and test wave front, $w_t(x, y)$, with the same wavelength, that are, respectively, described as

$$w_r(x, y) = a_r(x, y)e^{i[\Phi_r(x,y)]}, \tag{2.3}$$

$$w_t(x, y) = a_t(x, y)e^{i[\Phi_t(x,y)]}, \tag{2.4}$$

where $a_r(x, y)$ and $a_t(x, y)$ are the wave front amplitudes, and $\Phi_r(x, y)$ and $\Phi_t(x, y)$ are the wave front phases. When the reference and test wave fronts meet with each other, the equivalent wave front can be described as

$$w(x, y) = w_r(x, y) + w_t(x, y). \tag{2.5}$$

The intensity of the wave front, which represents the resultant interfering fringe pattern can be written as

$$I(x, y, t) = |w_r(x, y) + w_t(x, y)|^2, \tag{2.6}$$

or

$$I(x, y, t) = I'(x, y) + I''(x, y) \cos\left[\Phi_t(x, y) - \Phi_r(x, y)\right], \tag{2.7}$$

where

$$I'(x, y) = a_r^2(x, y) + a_t^2(x, y) \tag{2.8}$$

is the average intensity, and

$$I''(x, y) = 2a_r(x, y)a_t(x, y) \tag{2.9}$$

is the fringe or intensity modulation. If we define the phase difference as

$$\phi(x, y) = \Phi_t(x, y) - \Phi_r(x, y), \tag{2.10}$$

then we obtain the fundamental equation of the fringe analysis-based 3D shape measurement techniques:

$$I(x, y) = I'(x, y) + I''(x, y) \cos\left[\phi(x, y)\right], \tag{2.11}$$

where $I'(x, y)$ is the intensity bias, or DC component, $I''(x, y)$ is half of the peak-to-valley intensity modulation, and $\phi(x, y)$ is the unknown phase related to the temporal phase difference of this sinusoidal variation relative to the reference wave front. Obviously, if $\phi(x, y)$ here is $2k\pi(k = 0, 1, 2, \ldots)$, the interference pattern is at its peak when the travel distance between the reference wave front and the testing wave front is integer number of λ; and the interference pattern is at its valley when the travel distance is $(2n + 1)\lambda(n = 0, 1, 2)$.

If there is a temporal shift $\delta(t)$ in the testing wave front, the wave front phase at this location can be easily computed by the temporal phase shift, which is discussed in Section 2.3.2. The entire map of the unknown wave front phase $\phi(x, y)$ can be measured by monitoring and comparing this temporal shift at all measurement points, which is explained in the following section. From Equation 2.2, the distance between the reference wave front and the testing wave front can be mathematically determined as

$$h(x, y) = \frac{\phi(x, y)\lambda}{2\pi},\qquad(2.12)$$

assuming that the phase $\phi(x, y)$ obtained here does not have 2π ambiguity. It should be noted all these derivations are based on the assumption that the wave front phase can be directly measured. For reflection-based 3D shape measurement methods, where the light takes a round trip, the relationship between the surface height and the phase is scaled by $1/2$ as

$$h(x, y) = \frac{\phi(x, y)\lambda}{4\pi}.\qquad(2.13)$$

2.3 FRINGE ANALYSIS METHODS

2.3.1 Fourier Transform

To obtain phase information from the resultant fringe pattern, the Fourier method [27] can be employed. The general inference pattern equation, Equation 2.11, can be rewritten as

$$I(x, y) = I'(x, y) + \frac{1}{2}I''(x, y)\left[e^{j\phi(x,y)} + e^{-j\phi(x,y)}\right]\qquad(2.14)$$

in the complex form. In this equation, $e^{j\phi(x,y)}$ is the conjugate of $e^{-j\phi(x,y)}$. If a Fourier transform is applied, and the $I'(x, y)$, or the DC, and the conjugate components are filtered out by applying a band-pass filter, the recovered signal by an inverse Fourier transform leads to

$$\tilde{I}(x, y) = \frac{1}{2}I''(x, y)e^{j\phi(x,y)}.\qquad(2.15)$$

From this equation, the phase can be solved for by

$$\phi(x, y) = \tan^{-1}\left\{\frac{\Im[\tilde{I}(x, y)]}{\Re[\tilde{I}(x, y)]}\right\}.\qquad(2.16)$$

Here $\Im(\xi)$ and $\Re(\xi)$, respectively, denote the imagery and real part of ξ. The phase value, as a resultant of arctangent calculation, ranges from $-\pi$ to $+\pi$ with 2π modus. The phase map obtained here is also called the *wrapped* phase map. To obtain a continuous phase map, an additional step called phase unwrapping needs to be applied. The spatial phase-unwrapping step is essentially finding the 2π jumps from neighboring pixels, and removing them by adding or subtracting an integer number of 2π to the corresponding point [46]. The spatial phase-unwrapping algorithm is discussed in detail in Chapter 5. Besides spatial-phase unwrapping, there are also methods for temporal phase unwrapping by capturing more patterns, which is discussed in this chapter for the generic methods, and in Chapter 4 for the special methods developed for DFP systems. Finally, 3D coordinates can be reconstructed using the unwrapped phase, assuming that the system is properly calibrated [47]. Chapter 7 details the calibration methods developed for DFP systems.

As previously discussed, a single-fringe pattern can be used to recover a 3D shape. This method, as well as its variations, has been successfully adopted in numerous application areas [28], especially for extremely high-speed motion capture [45]. Yet, the success of Fourier method heavily relies on completely filtering out the DC and conjugate components, which may not be difficult for a uniform fringe image with approximately uniform surface reflectivity. However, the Fourier method may not work well if the surface geometry is complex or the surface reflectivity varies substantially from one point to the other. This is because the surface geometry or texture variation could substantially change the fringe pattern period, making the carrier fringe pattern spread around and thus difficult to be separated from the DC and conjugate components from the desired signal.

Since conventional Fourier method apply Fourier transform to the whole image, limiting its potential applications due the numerous problems including the aforementioned issues. To alleviate those problems, Qian [29,48] proposed the WFT method. Instead of applying global Fourier transform, WFT method applies Fourier transform to a local area that considers the local property of a fringe pattern and therefore, WFT is much more robust and substantially expands the applications of the single-fringe-based Fourier method. However, even with WFT, the influence of DC component cannot be ignored.

To further alleviate the problem associated with Fourier methods, one more pattern can be used to automatically eliminated the influence of DC component. One approach [30] is to directly project a uniform DC fringe pattern:

$$I_a(x, y) = I'(x, y). \tag{2.17}$$

For this method, before applying Fourier transform, the fringe image is obtained by taking the difference between the images,

$$I_m(x, y) = I(x, y) - I_a(x, y) = I''(x, y) \cos[\phi(x, y)]. \tag{2.18}$$

By doing so, the DC component does not significantly affect the measurement, thus the phase can be more accurately separated, improving measurement quality.

Since it is not easy for laser interferometry to generate uniform average intensity image, another variation [31] is to project the second sinusoidal pattern that has π phase shift, such that the difference between these two sinusoidal patterns cancel out the DC component, which gives the same benefit as the previous method. However, these methods sacrifice measurement speed because two fringe images must be captured to construct one 3D shape.

2.3.2 Phase Shifting

Equation 2.11 shows only three unknowns, $I'(x, y)$, $I''(x, y)$, and $\Phi(x, y)$, and thus only three equations are required to uniquely solve for all unknowns. Among multiple (three or more) fringe analysis techniques, the methods based on phase shifting have been extensively employed in optical metrology [49]. The basic concept behind phase-shifting method is that a time-varying phase shift is introduced between the reference wave front and the test or sample wave front in the interferometers. A time-varying signal $\delta(t)$ is then produced at each measurement point in the interferogram, and the relative phase between the two wave fronts at that location is encoded in these signals. The phase-shifting-based techniques have the following advantages:

- *The measurement speed could still be high.* This method only requires three fringe images to recover one 3D shape.

- *The spatial resolution is high.* This is because the phase can be obtained pixel by pixel, in other words, the measurement can be performed pixel by pixel.

- *It is robust to surface reflectivity variations.* The pixel-by-pixel phase calculation automatically cancels out the surface reflectivity for each point.

- *It is theoretically immune to ambient light.* The phase-shifting method also can automatically cancel out the DC component.

The general expression for the test wave fronts with a time-varying phase shift is

$$w_t(x, y, t) = a_t(x, y)e^{i[\Phi_t(x,y) - \delta(t)]}, \tag{2.19}$$

where $\delta(t)$ is the time-varying phase shift. When the reference described in Equation 2.3 and test wave front interfere with each other, the resultant intensity pattern is

$$I(x, y, t) = I'(x, y) + I''(x, y)\cos\left[\phi(x, y) + \delta(t)\right], \tag{2.20}$$

where $\delta(t)$, again, is the time-varying phase shift. The wave front phase at each point can be easily computed from this temporal delay. The entire map of the unknown wave front phase $\phi(x, y)$ can be measured by monitoring and comparing this temporal delay at all the required measurement points.

Over the years, there are many phase-shifting algorithms that have been developed to accomplish certain application needs. The selection of phase-shifting algorithm usually requires weigh the achievable speed, measurement accuracy, and the susceptibility to different error sources. Yet, there are a few well-established and extensively adopted phase-shifting algorithms that have been thoroughly described by Schreiber and Bruning [32] in Chapter 14 of the book edited by Dr. Malacara [49]. In this section, we only briefly summarize some of these algorithms, and we encourage the readers to refer the book by Dr. Malacara for detailed information.

2.3.2.1 Three-Step Phase-Shifting Algorithm

As aforementioned, three fringe images are the minimum number required to uniquely solve for the phase per pixel, and thus a three-step phase-shifting algorithm is desirable for high-speed 3D imaging applications. For a three-step phase-shifting algorithm [50] with the phase shifts of

$\delta(t) = -\alpha, 0, \alpha$, the mathematical descriptions of these three fringe images are

$$I_1(x, y) = I'(x, y) + I''(x, y) \cos[\phi(x, y) - \alpha], \tag{2.21}$$

$$I_2(x, y) = I'(x, y) + I''(x, y) \cos[\phi(x, y)], \tag{2.22}$$

$$I_3(x, y) = I'(x, y) + I''(x, y) \cos[\phi(x, y) + \alpha]. \tag{2.23}$$

Simultaneously solving these equations leads to

$$\phi(x, y) = \tan^{-1}\left[\frac{(1 - \cos\alpha)(I_1 - I_3)}{\sin\alpha(2I_2 - I_1 - I_3)}\right] \tag{2.24}$$

and

$$\gamma(x, y) = \frac{I''(x, y)}{I'(x, y)} = \frac{\sqrt{[(1 - \cos\alpha)(I_1 - I_3)]^2 + [\sin\alpha(2I_2 - I_1 - I_3)]^2}}{(I_1 + I_3 - 2I_2 \cos\alpha)\sin\alpha}, \tag{2.25}$$

or the data modulation indicating the fringe quality with 1 being the best. Data modulation is often valuable for phase unwrapping where the progress path is vital [46].

If the phase shift $\delta = 2\pi/3$, the phase and the data modulation can be, respectively, written as

$$\phi(x, y) = \tan^{-1}\left[\frac{\sqrt{3}(I_1 - I_3)}{2I_2 - I_1 - I_3}\right] \tag{2.26}$$

and

$$\gamma(x, y) = \frac{I''(x, y)}{I'(x, y)} = \frac{\sqrt{3(I_1 - I_3)^2 + (2I_2 - I_1 - I_3)^2}}{I_1 + I_2 + I_3}. \tag{2.27}$$

Three-step phase-shifting algorithm is good in terms of measurement speed since only three images are required; and it also has the merit of immune to $3k$ ($k = 1, 2, 3, \ldots$) harmonics error if the phase shift is $2\pi/3$ [51]. However, this method is quite sensitive to phase-shift error, which could be the most dominant error sources in laser interferometers. Moreover, this algorithm is quite sensitive to sensor noise (e.g., camera), since only three images are used and the averaging effect is not substantial.

One can easily derive that if $\delta = 2\pi/3$, the average intensity and the data modulation can be mathematically computed as

$$I'(x, y) = (I_1 + I_2 + I_3)/3, \tag{2.28}$$

$$I''(x, y) = \sqrt{3(I_1 - I_3)^2 + (2I_2 - I_1 - I_3)^2}/3. \tag{2.29}$$

Finally, the texture, a photograph of the image without fringe stripes, can be generated using

$$I_t(x, y) = I'(x, y) + I''(x, y). \tag{2.30}$$

2.3.2.2 Least-Squares Algorithm

Using more fringe patterns could be used to reduce noise influence or to alleviate the error caused by nonsinusoidality of the fringe pattern. In general, if more than three fringe images are used, the phase can be solved for by a least-square manner.

For an N-step phase-shifting algorithm [52,53], the intensity of the kth images with a phase shift of δ_k can be represented as

$$I_k(x, y) = I'(x, y) + I''(x, y) \cos(\phi(x, y) + \delta_k). \tag{2.31}$$

Simultaneously solving these N equations by a least-squares method leads to

$$\phi(x, y) = \tan^{-1}\left(\frac{-a_2(x, y)}{a_1(x, y)}\right) \tag{2.32}$$

and

$$\gamma(x, y) = \frac{I''(x, y)}{I'(x, y)} = \frac{[a_1(x, y)^2 + a_2(x, y)^2]^{1/2}}{a_0(x, y)}, \tag{2.33}$$

where

$$\begin{bmatrix} a_0(x, y) \\ a_1(x, y) \\ a_2(x, y) \end{bmatrix} = \mathbf{A}^{-1}(\delta_k)\mathbf{B}(x, y, \delta_k), \tag{2.34}$$

here

$$\mathbf{A}(\delta_k) = \begin{bmatrix} N & \sum \cos(\delta_k) & \sum \sin(\delta_k) \\ \sum \cos(\delta_k) & \sum \cos^2(\delta_k) & \sum \cos(\delta_k)\sin(\delta_k) \\ \sum \sin(\delta_k) & \sum \cos(\delta_k)\sin(\delta_k) & \sum \sin^2(\delta_k) \end{bmatrix} \tag{2.35}$$

and

$$\mathbf{B}(x, y, \delta_k) = \begin{bmatrix} \sum I_k \\ \sum I_k \cos(\delta_k) \\ \sum I_k \sin(\delta_k) \end{bmatrix}. \tag{2.36}$$

Here $I'(x, y) = a_0(x, y)$.

If the phase shift of the N-step phase-shifting algorithm is equal and in the form of

$$\delta_k = \frac{2k\pi}{N}, \quad k = 1, 2, 3, \ldots, N. \tag{2.37}$$

The phase equation can be simplified as

$$\phi(x, y) = -\tan^{-1}\left(\frac{\sum_{k=1}^{N} I_k \sin \delta_k}{\sum_{k=1}^{N} I_k \cos \delta_k}\right), \tag{2.38}$$

and the data modulation can be rewritten as

$$\gamma(x, y) = \frac{2\sqrt{\left(\sum_{k=1}^{N} I_k \cos \delta_k\right)^2 + \left(\sum_{k=1}^{N} I_k \sin \delta_k\right)^2}}{\sum_{k=1}^{N} I_k}. \tag{2.39}$$

In addition, the average intensity and intensity modulation can be derived as

$$I'(x, y) = \frac{\sum_{k=1}^{N} I_k}{N}, \tag{2.40}$$

$$I''(x, y) = \frac{2\sqrt{\left(\sum_{k=1}^{N} I_k \cos \delta_k\right)^2 + \left(\sum_{k=1}^{N} I_k \sin \delta_k\right)^2}}{N}. \tag{2.41}$$

In general, the more steps used, the better quality phase can be obtained, but the slower the measurement speed will be achieved.

2.3.2.3 Carré Algorithm
The aforementioned three-step phase-shifting and least-squares algorithms assume that the phase shift is precisely known. Unfortunately, depending upon the method adopted for phase-shift generation, the phase shift might not be precisely known, or might carry error; and the phase-shift

error creates problems for extracting precise phase values by adopting those phase shifting algorithms. To alleviate the problems associated with phase-shift error, Carré [54] proposed an algorithm that uses four fringe images with phase shifts of

$$\delta_k = -3\alpha, -\alpha, \alpha, 3\alpha, \quad k = 1, 2, 3, 4 \tag{2.42}$$

with the fringe images being described as

$$I_1(x, y) = I'(x, y) + I'' \cos[\phi(x, y) - 3\alpha], \tag{2.43}$$

$$I_2(x, y) = I'(x, y) + I'' \cos[\phi(x, y) - \alpha], \tag{2.44}$$

$$I_3(x, y) = I'(x, y) + I'' \cos[\phi(x, y) + \alpha], \tag{2.45}$$

$$I_4(x, y) = I'(x, y) + I'' \cos[\phi(x, y) + 3\alpha]. \tag{2.46}$$

This algorithm, instead of requiring the known phase shift values, directly computes the phase shift from these four fringe patterns as

$$\alpha = \tan^{-1} \sqrt{\left[\frac{3(I_2 - I_3) - I_1 + I_4}{I_1 + I_2 - I_3 - I_4} \right]} \tag{2.47}$$

and the phase

$$\phi = \tan^{-1} \left\{ \frac{\sqrt{[3(I_2 - I_3) - I_1 + I_4][I_1 + I_2 - I_3 - I_4]}}{-I_1 + I_2 + I_3 - I_4} \right\}. \tag{2.48}$$

This algorithm modestly compensates for phase-shift error, and even in the phase error varies spatially. However, this algorithm requires the phase shift for a given point to be equal. Furthermore, because the variation of phase shift, the conversion from phase to the height is no longer straightforward, albeit calibration approach could be developed if the phase shift error is systematic and constant.

2.3.2.4 Hariharan Algorithm

A better algorithm to tolerate large phase-shift error was proposed by Hariharan [55]. The idea behind this algorithm is to use five fringe patterns with phase shifts of

$$\delta_k = -2\alpha, -\alpha, 0, \alpha, 2\alpha, \quad k = 1, 2, 3, 4, 5 \tag{2.49}$$

and the fringe patterns are described as

$$I_1(x, y) = I'(x, y) + I'' \cos[\phi(x, y) - 2\alpha], \tag{2.50}$$

$$I_2(x, y) = I'(x, y) + I'' \cos[\phi(x, y) - \alpha], \tag{2.51}$$

$$I_3(x, y) = I'(x, y) + I'' \cos[\phi(x, y)], \tag{2.52}$$

$$I_4(x, y) = I'(x, y) + I'' \cos[\phi(x, y) + \alpha], \tag{2.53}$$

$$I_5(x, y) = I'(x, y) + I'' \cos[\phi(x, y) + 2\alpha]. \tag{2.54}$$

By minimizing the phase error using α as the variable, Hariharan found that the phase error is minimized when $\alpha = \pi/2$. Under optimal condition, the phase equation can be written as

$$\phi(x, y) = \tan^{-1}\left[\frac{2(I_2 - I_4)}{2I_3 - I_1 - I_5}\right] \tag{2.55}$$

and the data modulation is

$$\gamma(x, y) = \frac{3\sqrt{4(I_4 - I_2)^2 + (I_1 + I_5 - 2I_3)^2}}{2(I_1 + I_2 + 2I_3 + I_4 + I_5)}. \tag{2.56}$$

This algorithm can tolerate large phase-shift error even though it requires five fringe images and slows down the measurement speed.

2.4 TWO- AND MULTIWAVELENGTH PHASE-SHIFTING ALGORITHMS

As aforementioned, the phase obtained from a single-wavelength phase-shifting method is within the range of $[-\pi, \pi)$. When a fringe pattern contains more than one periodical fringe stripes, the phase need to be unwrapped to obtain a continuous phase map. This means that if another set of wider fringe patterns (e.g., a single fringe stripe can cover the whole image) is used to obtain a phase map without 2π discontinuities. The second phase map can be used to unwrap the narrow wrapped phase point by point without spatially unwrapping the phase. To obtain the phase of wider fringe patterns, there are two approaches: (1) use a very long wavelength directly and (2) use two or more short wavelengths [56,57] to generate an equivalent long wavelength. The former is not very commonly used because it is difficult to generate high-quality wide fringe patterns due to noises or hardware limitations. Thus, the latter is more frequently adopted. This section briefly explains the principle of this technique.

Theoretically, the relationship between the phase Φ and the wavelength λ, and the height $h(x, y)$ can be written as Equation 2.2 regardless of the wavelength used. If two different wavelengths (λ_1, λ_2 and $\lambda_1 < \lambda_2$) are used to measure the same object surface, two phase maps can be obtained:

$$\Phi_1 = \Phi(x, y) = \frac{2\pi h(x, y)}{\lambda_1}, \tag{2.57}$$

$$\Phi_2 = \Phi(x, y) = \frac{2\pi h(x, y)}{\lambda_2}. \tag{2.58}$$

Taking the difference of these two phase maps leads to

$$\Delta\Phi_{12} = \Phi_1 - \Phi_2 = [2\pi \cdot h(x, y)/\lambda_{12}^{eq}] \tag{2.59}$$

Here, $\lambda_{12}^{eq} = \lambda_1\lambda_2/|\lambda_2 - \lambda_1|$ is the equivalent wavelength between λ_1 and λ_2. If $\lambda_2 \in (\lambda_1, 2\lambda_1)$, we have $\lambda_{12}^{eq} > \lambda_2$. In reality, we only have the wrapped phase, ϕ_1 and ϕ_2. We know that the relationship between the absolute phase is Φ and the wrapped phase

$$\phi = \Phi \quad (\text{mod } 2\pi)$$

with 2π discontinuities. Here the modulus operator is to convert the phase to a range of $[0, 2\pi)$. Taking the modulus operation on Equation 2.59 leads to

$$\Delta\phi_{12} = [\Phi_1 - \Phi_2] \quad (\text{mod } 2\pi) = [\phi_1 - \phi_2] \quad (\text{mod } 2\pi). \tag{2.60}$$

$\Delta\phi_{12} = \Delta\Phi_{12} \ (\text{mod } 2\pi)$. If the wavelengths are properly chosen, so that the resultant equivalent wavelength λ_{12}^{eq} is large enough to cover the whole range of image. In this case, the modulus operator does not change the phase, thus no phase unwrapping is required.

However, because the existence of noise, two-wavelength technique is usually not sufficient [58]. Practically, three or more wavelengths are required for point by point absolute phase measurement. The multiwavelength technique is designed so that the equivalent widest fringe stripe can cover the whole image [59].

Assume another set of fringe patterns with the wavelength (λ_3) are used, the equivalent wavelength between λ_1 and λ_3 is $\lambda_{13}^{eq} = \lambda_1\lambda_3/|\lambda_3 - \lambda_1|$

and we have

$$\Delta\phi_{13} = [\phi_1 - \phi_3] \quad (\text{mod } 2\pi), \tag{2.61}$$

$$\Delta\phi_{123} = (\Delta\phi_{13} - \Delta\phi_{12}) \quad (\text{mod } 2\pi). \tag{2.62}$$

Here, $\lambda_{123}^{eq} = \lambda_{12}^{eq}\lambda_{13}^{eq}/|\lambda_{13}^{eq} - \lambda_{12}^{eq}|$. Once the absolute phase of the longest equivalent wavelength is obtained, one can reversely unwrap the phase of other wavelengths. The phase of the shortest wavelength is usually used to recover 3D information because the measurement accuracy is approximately inversely proportional to the wavelength due to the presence of noise.

2.5 SUMMARY REMARKS

This chapter explained the fundamental theory behind fringe generation, which is based on wave optics. Various fringe analysis techniques including the phase-shifting methods were discussed. The multiwavelength phase-shifting algorithms were presented as the last part of this chapter. All algorithms or approaches discussed in this chapter are applicable to any 3D shape measurement techniques including laser interferometry and DFP technique. The next chapter presents the digital fringe generation technique, and discusses the differences among three extensively adopted digital video projection technologies, namely, DLP, LCD, and LCoS technologies.

Digital Fringe Generation Technique

D UE TO ITS ACHIEVABLE SPEEDS, simple system setup, and digital pattern generation nature, the DFP techniques play an increasingly important role in 3D shape measurement in the past decade. Instead of generating sinusoidal patterns by interference, the DFP technique directly projects sinusoidal patterns onto object surfaces without interference. Since DFP technique allows the use of wide spectrum of light (e.g., white light), the speckle noise pertinent to laser system does not present. This chapter explains the basic principle behind DFP system and presents the differences between three different types of digital video projection technologies, namely DLP, LCD, and LCoS.

3.1 INTRODUCTION

The advantages of the DFP technique over other techniques for 3D shape measurement mainly due to its *digital* fringe generation nature. Comparing to analog fringe generation methods (e.g., grating), the DFP methods tend to be more flexible, easier, and quicker; and comparing to interference-based fringe generation method, DFP methods do not suffer from speckle noise and phase-shift error. However, DFP methods could produce undesirable artifacts that compromises fringe quality. To generate high-quality sinusoidal patterns, the fringe projection device (e.g., digital video projector) itself needs to be fully understood, and sometimes even the interface between a computer and a digital fringe projection device (e.g., digital video card) also needs to be carefully handled. Since modern interface hardware is quite robust and consistent, its influence can be ignored. Therefore, we only consider how the digital projection device generates fringe patterns.

There are three major digital video projection technologies, DLP (Digital Light Processing), LCD (Liquid Crystal Display), and LCoS (Liquid Crystal on Silicon).

3.2 BASICS OF DFP SYSTEM

As discussed in Section 1.5 of Chapter 1, the DFP system uses a computer to generate sinusoidal fringe patterns that are directly projected onto the object surface. The distorted fringe patterns are then captured by the camera viewing from a different angle. By analyzing the fringe pattern distortion in phase domain, 3D information can be recovered using geometric triangulation. The key to the success of a DFP technique is to generate high-quality fringe patterns. Ideally, assuming the projection system has linear response to input, the computer generated fringe pattern can be mathematically described as

$$I(i,j) = 255/2[1 + \cos(2\pi j/P + \delta)]. \tag{3.1}$$

Here, P is fringe pitch, or the number of pixels per fringe period, δ is the phase shift, (i, j) is the pixel index, and the pattern varies sinusoidally along j direction. It should be noted that the commercially available projectors usually are nonlinear, which require nonlinearity compensation. And the nonlinearity calibration and compensation method are discussed in Chapter 6. In this chapter, we assume that the projector is an ideally linear system that can project grayscale values from 0 to 255 to simplify conceptual discussions and mathematical representations.

For example, three phase-shifted fringe patterns with equal phase shifts can be represented as

$$I_1(i,j) = 255/2[1 + \cos(2\pi j/P - 2\pi/3)], \tag{3.2}$$

$$I_2(i,j) = 255/2[1 + \cos(2\pi j/P)], \tag{3.3}$$

$$I_3(i,j) = 255/2[1 + \cos(2\pi j/P + 2\pi/3)]. \tag{3.4}$$

For the computer-generated fringe patterns, phase shifts can be directly introduced in spatial fringe patterns rather than by time domain (as those discussed in Chapter 2).

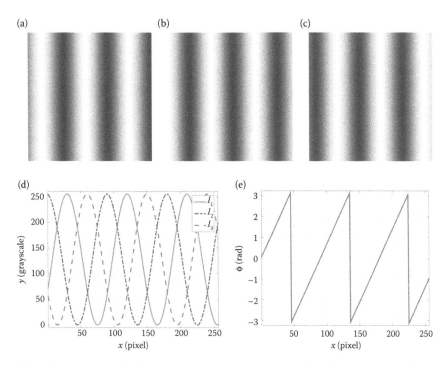

FIGURE 3.1 Example fringe pattern generation and the phase. (a)–(c) Three phase-shifted fringe patterns with equal spatial phase shifts. (d) Cross sections of these fringe patterns. (e) Cross section of the wrapped phase map.

Figure 3.1a–c show three phase-shifted fringe patterns with a fringe pitch of 90 pixels, and an image resolution of 256×256. These patterns are spatial shifted by $2\pi/3$, or $1/3$ of the fringe periods (30 pixels) from one pattern to the next. Figure 3.1d shows the cross section of these phase-shifted patterns, from which the wrapped phase can be computed, as shown in Figure 3.1e.

3.3 DIGITAL VIDEO PROJECTION TECHNOLOGIES

Once the patterns are generated by a computer, they can be sent to a digital video projector for projection. Commercially, there are three major types of video projection technologies that can be used for DFP systems: DLP, LCD, and LCoS technologies. Understanding the different principles behind these technologies allows one to select the best technology for a particular system design. In this section, we briefly describe the operation mechanism of each technology and discuss their advantages and shortcomings comparing with the other.

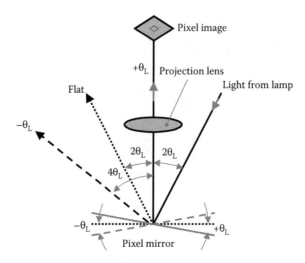

FIGURE 3.2 Optical switching principle of a digital micromirror device (DMD).

3.3.1 Digital Light Processing

DLP concept originated from Texas Instruments (TI) in the late 1980s. In 1996, TI began commercializing its DLP technology. At the core of every DLP projection system, there is an optical semiconductor called the digital micromirror device, or DMD, which functions as an extremely precise light switch. The DMD chip contains an array of hinged, microscopic mirrors, each of which corresponds to one pixel of light in a projected image.

Figure 3.2 shows the working principle of the micromirror. Data in the cell control electrostatic forces that can move the mirror $+\theta_L$ (ON) or $-\theta_L$ (OFF), thereby modulating the light that is incident on the mirror. The rate of a mirror switching ON and OFF determines the brightness of the projected image pixel. An image is created by light reflected from the ON mirrors passing through a projection lens onto a screen. Grayscale values are created by controlling the proportion of ON and OFF times of the mirror during one frame period—black being 0% ON time and white being 100% ON time.

DLP projectors embraced the DMD technology to generate color images. All DLP projectors include a light source, a color filter system, at least one DMD, DLP electronics, and an optical projection lens. For a single-chip DLP projector, illustrated in Figure 3.3, the color image is produced by placing a color wheel into the system. The color wheel, that contains red, green, and blue filters, spins at a very fast speed, thus red, green, and blue channel images will be projected sequentially onto the screen. However, because the

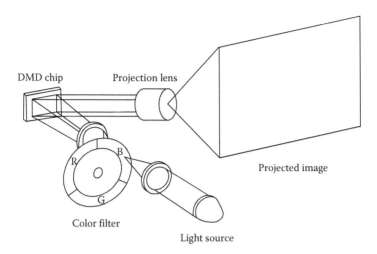

FIGURE 3.3 Configuration of a single-chip digital line processing (DLP) projector. (From P. S. Huang, C. Zhang, and F.-P. Chiang, *Opt. Eng.* 42, 163–168, 2003. With permission.)

refresh rate is so high, human eyes can only perceive a single color image instead of three sequential ones.

A DLP projector produces a grayscale value by time integration [60]. A simple test was performed for a single-chip DLP projector (BenQ W770ST, Taipei, Taiwan) that runs at a projection speed of 120 Hz. The output light was sensed by a photodiode (Thorlabs FDS100, Newton, New Jersey), and the photocurrent was converted to voltage signal and monitored by an oscilloscope. The photodiode used has a response time of 10 ns, an active area of 3.6 mm × 3.6 mm, and a bandwidth of 35 MHz. The oscilloscope used to monitor the signal was Tektronix TDS2024B (Beaverton, Oregon) and has a bandwidth of 200 MHz.

Figure 3.4 shows some typical results when the DLP projector was fed with uniform images with different grayscale values. If the pure green, RGB = (0, 255, 0), is supplied, the signal has the duty cycle of almost 100% ON. When the grayscale value is reduced to 128, approximately half of the channel is filled. If the input grayscale value is reduced to 64, a smaller portion of the channel is filled. If the input grayscale value is 0, the signal has the duty cycle of almost 0% ON. These experiments show that if the supplied grayscale value is somewhere between 0 and 255, the output signal becomes irregular. Therefore, if a sinusoidal fringe pattern varying from 0 to 255 is supplied, the whole projection period must be used to correctly represent the image projected from the projector.

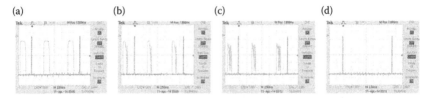

FIGURE 3.4 Example of the digital line processing (DLP) projected timing signal if the projector is fed with different grayscale value of the green image. (a) Green = 255; (b) green = 128; (c) green = 64; (d) green = 0.

3.3.2 Liquid Crystal Display

LCD concept was discovered in the eighteenth century and made commercially successful in the 1990s. Figure 3.5a illustrates the operation principle. The entering light is first polarized by a vertical polarizing filter (P1); and the polarized light passes through the negative electrode layer (E1), enters the liquid crystal (LC) layer, passes through the positive electrode layer (E2), and then enters into the horizontal polarizing filter (P2). By changing the voltage applied to those two electrodes (E1 and E2), the crystal status will be modulated, changing the percentage of light that will pass through the polarizing filter P2, which modulates the light intensity and thus represents the grayscale values. If there is no voltage applied to electrodes, theoretically, 0% of light will pass through the LCD, and thus the grayscale value will be 0; if the highest voltage is applied to electrodes, theoretically, 100% of light will pass through the LCD, representing grayscale value of 255; and intermediate grayscale values will be generated by changing the voltage. By this means, the LCD can display any grayscale values instantaneously (with a short response time taking by the crystal to change its status).

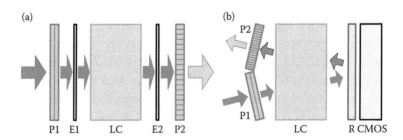

FIGURE 3.5 Optical principle of (a) liquid crystal display (LCD) and (b) liquid crystal on silicon (LCoS) display.

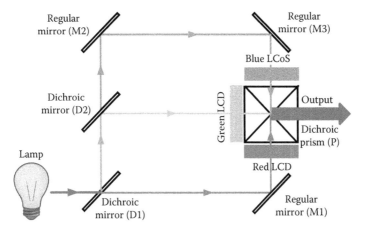

FIGURE 3.6 **(See color insert.)** Possible configuration of a liquid crystal display (LCD) projector.

Figure 3.6 illustrates one possible optical configuration for an LCD projection system. The white light coming out of a lamp is split into two different color spectral, the red and the blue–green, by a dichromic mirror (D1). The red light, redirected by a regular mirror (M1), passes through the red LCD, and enters the dichroic prism (P). The blue–green light is split again by another dichroic mirror (D2) into blue and green lights. The green light eventually enters the dichroic prism after passing through the green LCD. Similarly, the blue light enters the dichroic prism after being redirected by two regular mirrors (M2 and M3) and passing through the blue LCoS. Finally, the dichroic prism combines all the lights that are sent to the projection lens for display.

The same photodiode was used to test the response of an LCD projector (Epson 730HD, Nagano, Japan) to different grayscale images. Again, uniform green images with grayscale values of 255, 128, 64, and 0 were tested. Figure 3.7 shows the result. These experimental results demonstrated that for a given input grayscale value, the output light intensity is rather stable for the whole projection period, which is different from DLP (no gap between pulses); and the intensity of light varies overall corresponding to the input grayscale values.

3.3.3 Liquid Crystal on Silicon

LCoS display was first demonstrated by General Electric (GE) in the late 1970s; and a number of companies attempted to develop LCoS projectors in the 1990s. LCoS is a hybrid technology that uses the basic principle of

(a)   (b)   (c)   (d)

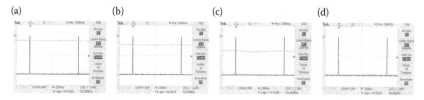

FIGURE 3.7   Example of the liquid crystal display (LCD) projected timing signal if the projector is fed with different grayscale values of the green image. (a) Green = 255; (b) green = 128; (c) green = 64; (d) green = 0.

LCD technology to modulate light intensity by controlling LCs, but uses DLP technology's reflective rather than LCD's transmissive technology to direct the light. Figure 3.5b illustrates the operation principle. The entering light is firstly polarized by a vertical polarizing filter (P1); and the polarized light passes through the LC layer to reach the reflective coating layer (R) that bounces back the light. The silicon layer controlling the LC with a chip or sensor is a chip or sensor, a complementary metal-oxide-semiconductor (CMOS). The reflected light passes through the LC again to reach another polarizing filter (P2) whose polarizing angle is perpendicular to P1, or horizontal. Similar to LCD, different grayscale values are generated by changing the LC status. Because of the similarity between the LCoS and LCD, LCoS is often regarded as reflective LCD, instead of transmissive, LCoS is reflective. Comparing with LCD or DLP, LCoS tends to start with higher resolution, albeit lower resolution LCoS systems are also available on the market.

Figure 3.8 illustrates one possible optical configuration for an LCoS projection system. The white light coming out of a lamp is split into two different color spectral, the red and the blue–green, by a dichromic mirror (D1). The red light enters a polarizing beam splitter (B1), hits the red LCoS, and is reflected back to the dichroic prism (P). The blue–green light, redirected by a regular mirror (M1), is split again by another dichroic mirror (D2) into blue and green lights. The green light eventually enters the dichroic prism after passing through the polarizing beam splitter (B2) and being reflected back by the green LCoS. Similarly, the blue light enters the dichroic prism after passing through the polarizing filter (B3) and being reflected by the blue LCoS. Finally, the dichroic prism combines all the lights that are sent to the projection lens for display.

Similar experiments were carried out to demonstrate the response of an LCoS projector (Canon Realis SX50, Tokyo, Japan) to different grayscale images. Again, uniform green images with grayscale values of 255, 128, 64,

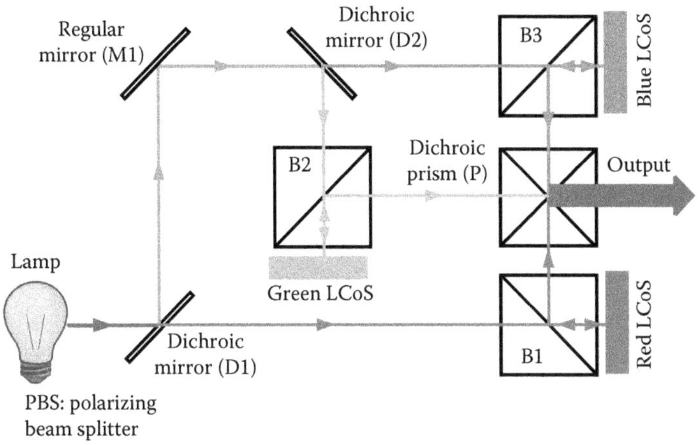

Regular
mirror (M1)

Dichroic
mirror (D2)

B3

Blue LCoS

B2

Dichroic
prism (P)

Output

Lamp

Green LCoS

Dichroic
mirror (D1)

B1

Red LCoS

PBS: polarizing
beam splitter

FIGURE 3.8 **(See color insert.)** Possible configuration of a liquid crystal on silicon (LCoS) projector.

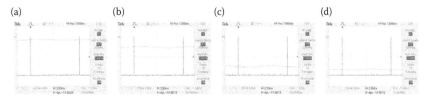

FIGURE 3.9 Example of the liquid crystal on silicon (LCoS) projected timing signal if the projector is fed with different grayscale values of the green image. (a) Green = 255; (b) green = 128; (c) green = 64; (d) green = 0.

and 0 were used for testing. Figure 3.9 shows the result. These experimental results demonstrated that similar to LCD but different from DLP (no gap between pulses); and the intensity of light varies overall corresponding to the input grayscale values.

3.3.4 Practical Issues for Using Digital Video Projection Technologies

For 3D shape measurement with a DFP method, precisely capturing high-quality sinusoidal fringe patterns is the key to achieving high measurement accuracy. Due to the differences between different projection technologies, the flexibility of system development, and the ultimate achievable accuracy could be different. The major issues to consider include

1. *Image contrast.* Image contrast is vital for high-accuracy 3D shape measurement, since DFP systems use sinusoidal patterns that vary from 0 to 255. The higher contrast the projector can generate, the better signal-to-noise ratio (SNR) the system can produce, and thus the better resolution the system can achieve. Theoretically, single-chip DLP projectors provide higher contrast than LCoS or LCD. This is because a single-chip DLP projection system can completely turn ON and OFF. Furthermore, DLP offers some unique features such as high-speed binary image switching (at tens of kHz). This is of great interest to some of our recent research on superfast 3D shape measurement based on the binary defocusing method, which is not the subject of this book; and if interested, one can read the summary paper [61]. Comparing with LCD, LCoS could have higher contrast because its filling factor is higher (or gap between pixels is smaller). It is important to note that with the advancement of fabrication technologies, the differences between LCD and LCoS are smaller and smaller.

2. *Grayscale generation.* From prior discussions, DLP generates grayscale values through time integration that utilizes the full projection cycle to represent the desired grayscale image. In contrast, LCD or LCoS generates the grayscale values by modulating the crystal status, which occurs instantaneously without time integration; and thus any segment of the projection cycle could be used to represent the desired grayscale image.

3. *Color generation.* It is easy to understand that for those color-coded DFP systems, the color plays a critical role. However, because of the problems associated with color, the white light is often preferable for high-quality 3D shape measurement. Unlike LCoS or LCD projectors where three chips are typically utilized to generate color simultaneously, the extensively adopted DLP projector uses a single-chip DMD, and a rapidly spinning color wheel to generate color sequentially. If not properly handled, the color introduced artifacts may not be negligible. For example, Exposures 1 and 2 illustrated in Figure 3.10 have the same duration, the camera-captured images are identical if an LCoS or LCD is used. However, the camera images are different if a single-chip DLP projector is used. This is because a typically CCD or CMOS camera sensor sensitivity is different for different colors. It should be noted that the special projection mechanism of a DLP system is not always a shortcoming: Zhang and Huang have taken

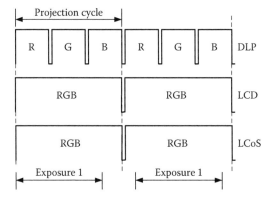

FIGURE 3.10 The mechanism of generating color images could lead to measurement error for single-chip digital line processing (DLP)-based systems. DLP generates three primary colors in sequence, but liquid crystal display (LCD) and liquid crystal on silicon (LCoS) typically generate three colors simultaneously.

advantage of this unique projection mechanism to develop some of the real-time 3D imaging technologies [62]. Despite these differences, it is always safe to expose the camera with one or multiple full projection period if the measurement speed is not the top priority.

4. *Synchronization requirement.* The less stringent synchronization requirement between the camera and the projector, the more flexible the DFP system will be. For 3D imaging, especially high-speed 3D imaging, the projector and the camera require a certain level of synchronization requirement because multiple patterns are switched. Comparing with DLP, LCoS, or LCD has the advantage of generating the desired grayscale values instantaneously rather than through time integration. Therefore, for a DFP system, using an LCoS or LCD projector is not as sensitive to projector–camera synchronization as using a DLP projector [63]. Furthermore, shorter exposure time than the refreshing period can be used if an LCoS or LCD projector is used, making the system development more flexible than that using a DLP projector.

It is important to note that the discussions presented in this section is true only when the grayscale sinusoidal fringe patterns are sent to the projector and let the projector naturally generates sinusoidal patterns. The recent study shows that if the binary defocusing technology [64] is adopted, the problems associated with the timing for the DLP technology no longer present, and the high-contrast and high-speed nature of DLP becomes a prominent choice for 3D imaging system development [61].

3.3.5 Comparing Experimental Results

To find the differences between projection technologies, three different types of projectors were tested. Table 3.1 shows the detailed specifications of these projectors.

For all experiments, the same camera (Jai Pulnix TM-6740CL, San Jose, California) was used. The camera has a resolution of 640 × 480 and a maximum frame rate of 200 frames/s. The pixel size of the camera is $7.4 \times 7.4 \ \mu m^2$. The camera was used with a 16-mm focal length megapixel lens (Computer M1614-MP) at F/1.4 to 16C. To fairly compare them, the projectors should have the same amount of focusing, the captured fringe periods were the same, and the projector illumination areas were similar.

As discussed above, the single-chip DLP projector generates grayscale images in a completely different manner from the LCD and the LCoS projectors. The particular DLP projector used in this study generates two pulses in a 1/60-s projection cycle for one color channel (e.g., green), while the LCD and the LCoS projector stay ON through the projection cycle regardless of the input grayscale values. Since the LCD and the LCoS do not generate any pulses during projection, the timing used for the DLP projector is shared with the other two projectors. The camera exposure was set to start with the computer's VSync signal and end either on a DLP pulse (the middle of the pulse) or off a pulse (the middle between two pulses) [65]. For green channel, five exposure times were selected, they are 5.70, 10.00, 14.10, 16.00, and 16.67 ms. Table 3.2 summarizes the exposure times used for each color channel, and the white channel uses all exposure time combination.

The performance of each individual projector was examined with the predetermined camera exposure times. A uniform white flat board was measured to evaluate the performance for all three projectors. In these measurements, the phase was captured using a three-step phase-shifting algorithm and a temporal phase-unwrapping method was used to obtain

TABLE 3.1   Projectors Specifications

Projection	Resolution	Contrast	Output (lm)	Aperture	Focal Length (mm)	Chip No.
DLP	1280 × 720	13,000 : 1	2500	F/2.60–2.78	10.20–12.24	Single
LCD	1400 × 1050	12,000 : 1	3000	F/1.58–1.72	16.90–20.28	Three
LCoS	1280 × 800	1000 : 1	2500	F/1.85–2.5	22.00–37.00	Three

TABLE 3.2   Exposure Times Used for Experimentations

Color Channel	Exposure 1 (ms)	Exposure 2 (ms)	Exposure 3 (ms)	Exposure 4 (ms)	Exposure 5 (ms)
Red	7.60	11.80	15.90	16.67	
Green	5.70	10.00	14.10	15.90	16.67
Blue	3.50	7.50	11.80	15.00	16.67

Note: Note that red channel only uses four exposures instead of five exposures because the projector ends with red channel per projection cycle.

the continuous phase map Φ. The phase error was determined by comparing against the *ideal* phase, Φ^i, captured using narrow fringe patterns with a nine-step phase-shifting method and a binary defocusing technique [66]. Φ^i was also smoothed by a Gaussian filter to reduce the random noise effect. Thus, $\Delta\Phi = \Phi - \Phi^i$ indicates the phase error, and rms error of $\Delta\Phi$ was calculated to represent the phase error.

Figure 3.11 shows the measurement results. All experiments shown in this section were captured using patterns with a fringe period of 18 pixels; the nonlinear gamma effect was also precorrected for all projectors using the method to be discussed in Chapter 6. In order to capture data with reasonable quality for a large camera exposure range, the first set of experiments was performed by adjusting camera aperture manually. The brightness of the image was visually inspected so that they are approximately the same. From the experimental data, it can be seen that when the exposure time was a full projection cycle (i.e., exposure time of 16.67 ms), the DLP projector always provided the smallest phase error, while the LCoS projector always produces the largest error. This is due to the lowest contrast

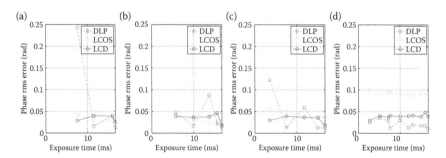

FIGURE 3.11 **(See color insert.)** Comparison between liquid crystal on silicon (LCoS) and digital line processing (DLP) projectors when camera aperture was manually adjusted. These figures show the phase rms errors. (a) Red channel; (b) green channel; (c) blue channel; (d) white light.

level of the LCoS projector compared with the other two projectors. While the difference in performance at full cycle is not significant when using single color channels, the differences become more noticeable when using the white light. When a exposure time is shorter than one projection cycle, it can be noticed that the DLP's performance is not stable, the ON pulse results are more than the OFF pulse results. In contrast, LCD and LCoS's performances are relatively stable through the projection cycle. Their performances are similar using single color channels. However, when using the white light, the LCD projector can provide significantly better phase quality than the LCoS projector does.

As shown in Figure 3.11a–c, for the DLP projector, the phase error using the red channel is the largest. This is probably because a single DLP projector projects different color channels sequentially; the DLP projector starts with the blue channel, then the green and red channels in sequence. If the red channel is used, the projector will have a longer off time before the projected signal comes. This means that more ambient light will enter the camera, lower the SNR and thus the captured phase quality will be affected.

Since the first set of experiments was performed by adjusting the camera aperture manually, though visually inspected, the differences in brightness between measurements are inevitable. This could be the reason that the LCD and the LCoS's error curves shown in Figure 3.11 are not smooth. To reduce the uncertainty caused by visual inspection, a second set of measurements was carried when the camera's aperture is not adjusted. Table 3.3 shows the results. This again verifies that the DLP's performance significantly varies from ON pulse to OFF pulse; while LCD and LCoS's performances barely have any differences.

To demonstrate the differences between the three technologies more intuitively, a more complex 3D object was measured. The green channel was used along with two exposure times: 5.70 ms (ON pulse exposure time for the DLP) and 10.00 ms (OFF pulse exposure time for the DLP). The nonlinear gamma was, again, corrected for all projectors before

TABLE 3.3   Phase rms Error When Only Green Channel Was Used with Exposure Time of 5.70 ms (ON Pulse) and 10.00 ms (OFF Pulse)

	ON Pulse	OFF Pulse
DLP (rad)	0.0496	0.0169
LCD (rad)	0.0386	0.0364
LCoS (rad)	0.0370	0.0356

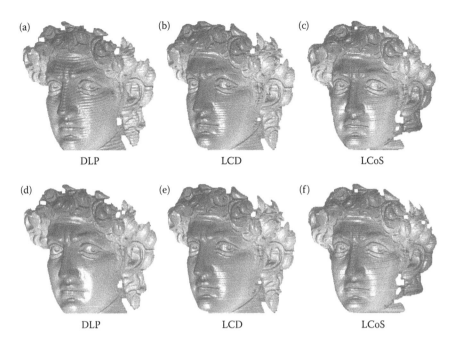

FIGURE 3.12 Complex 3D object measurement results using sinusoidal digital fringe projection method when the camera is synchronized with the projector. (a–c) ON pulse (5.70 ms) results; (d–f) corresponding OFF pulse (10.00 ms) results.

experiments. The sample phase-to-height conversion method introduced in Reference 67 was used to convert the unwrapped phase to 3D shape. Figure 3.12 shows the measurement results. It is clear that DLP can generate the best results using the OFF pulse exposure, but it produces the worst results when the ON pulse exposure was used. In contrast, the LCD and the LCoS have much stable performances: though their results cannot match DLP's when OFF pulse exposure was used, they provided better results when ON pulse exposure was used. Comparing the LCD and the LCoS, one may also notice that the LCD projector offers better results, this could be a result of its higher contrast.

As noted above, if the binary defocusing method [64] is used, DLP no longer has some limitations. Figure 3.13 shows comparing results using the same exposure time as the ones presented in Figure 3.12. All projectors were properly defocused before any measurements. Results clearly show that the DLP projector consistently provides the highest quality phase. Maybe because of irregular pixels, the results from both the LCD and the LCoS projectors have some structure error. And probably due to their lower contrast compared to the DLP, they also have more random noise.

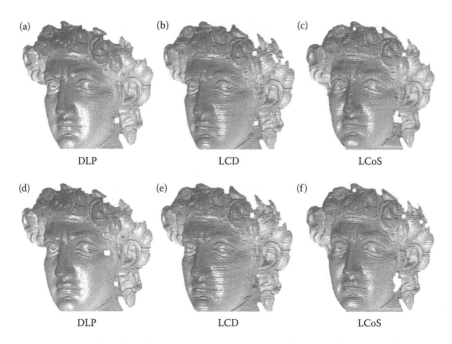

FIGURE 3.13 Complex 3D object measurement results using the binary defocusing method when the camera is synchronized with the projector. (a–c) ON pulse (5.70 ms) results; (d–f) corresponding OFF pulse (10.00 ms) results.

3.4 SUMMARY REMARKS

DFP technologies has its overwhelming advantages over other 3D imaging techniques because its achievable high accuracy, high speed, and flexibility. This chapter summarized three key digital video projection technologies: DLP, LCD, and LCoS. Because of their different operation mechanisms, measurement error could be introduced without careful considerations and handling. The key to successfully perform measurements for a DFP system is to properly generate fringe patterns with a computer, send those patterns to the projector, and then capture those patterns by the camera. It is highly desirable to capture high-quality sinusoidal for high-quality 3D imaging. Yet, in most cases, before any 3D imaging, phase unwrapping is usually required. Phase unwrapping essentially convert phase maps with 2π discontinuities to smooth phase maps before they can be used for 3D reconstruction. Therefore, phase unwrapping plays a key role in 3D imaging. In the following two chapters, we discuss two types of phase-unwrapping algorithms: temporal phase-unwrapping algorithms and spatial phase-unwrapping algorithms.

CHAPTER 4

Temporal Phase Unwrapping for Digital Fringe Projection Systems

T HE PHASE OBTAINED FROM a single-wavelength fringe pattern(s) only provides phase values ranging $(-\pi, +\pi)$ with 2π discontinuities. For 3D imaging, continuous phase is required, and thus properly removing 2π jumps is necessary. Multiwavelength phase-shifting algorithms introduced in Chapter 2 is one approach for laser-interference-based systems, and the multiwavelength phase-shifting algorithms can be adapted to the DFP systems. Due to the versatility of DFP systems, more temporal phase-unwrapping methods can be realized that are not possible for laser interference systems. This chapter summarizes some major temporal phase-unwrapping algorithms that can be employed for DFP systems.

4.1 INTRODUCTION

Ultimately, phase unwrapping is to determine 2π discontinuous locations, and then remove them by adding or subtracting multiple number of 2π to the wrapped phase, $\phi(x, y)$. In other words, the relationship between the wrapped phase and the unwrapped phase, $\Phi(x, y)$ can be mathematically represented as

$$\Phi(x, y) = k(x, y) \times 2\pi + \phi(x, y). \tag{4.1}$$

Here $k(x, y)$ is an integer numbers that represents fringe order. If $k(x, y)$ can be uniquely determined for each fringe stripe, then $\Phi(x, y)$ can be regarded as *absolute* phase. Temporal phase unwrapping typically provides absolute phase. In contrast, the spatial phase unwrapping, to be discussed in Chapter 5, can determine the integer number $k(x, y)$ with an offset, the unwrapped phase is called *relative* phase, but not *absolute* phase. The fundamental difference between a spatial phase unwrapping and a temporal phase unwrapping is that the temporal phase unwrapping does not require to know the neighboring pixel phase information to perform phase unwrapping; and thus the temporal phase unwrapping is suitable for measuring arbitrary geometric shape objects. In contrast, the spatial phase-unwrapping algorithm hinges on detecting 2π discontinuities from neighboring pixels; and thus it requires the surface to be *smooth* (e.g., surface geometry cannot induce 2π phase changes). Therefore, if the spatial phase-unwrapping algorithm is typically limited to measure *smooth* surfaces without isolated islands.

The following sections discuss some of the major temporal phase-unwrapping algorithms that are applicable to the DFP systems.

4.2 MULTIFREQUENCY PHASE-SHIFTING ALGORITHM

The multifrequency phase-shifting algorithms are based on the multiwavelength phase-shifting algorithms discussed in Chapter 2. Here the fringe frequency f or fringe period T makes more sense since the sinusoidal fringe patterns are directly generated by computer. If the fringe period T is used, all derivation of the multifrequency phase-shifting algorithms are identical to those multiwavelength phase-shifting algorithms. In other words, the equivalent fringe period T^{eq} can be written as

$$T^{eq} = \frac{T_1 T_2}{|T_1 - T2|}, \tag{4.2}$$

where T_1 and T_2 are fringe periods for two higher frequency fringe patterns. Similarly, equivalent phase can be obtained for ϕ_1 and ϕ_2 as

$$\phi_{12} = [\phi_1 - \phi_2] \mod (2\pi). \tag{4.3}$$

By using multiple-frequency fringe patterns, the overall equivalent fringe period can cover the whole projection range, making the equivalent phase to be the *absolute* phase

$$\Phi(x, y) = \phi_{eq}(x, y). \tag{4.4}$$

For a DFP method, fringe periods are usually in pixels and have the freedom to choose any fringe periods as long as they are integer numbers (here integer numbers are used to avoid subpixel fringe period representation error). The ultimate goal of utilizing a multi-frequency phase-shifting algorithm for temporal phase unwrapping is to generate the longest equivalent fringe period larger than the pattern span. For example, to cover the span of fringe pattern size of 320×320, one can chose a three-frequency phase-shifting algorithm with fringe periods of $T_1 = 30$, $T_2 = 45$, and $T_3 = 72$ pixels, resulting in an equivalent fringe period of 360 pixels. Figure 4.1 illustrates the procedures of generating the equivalent fringe period of 360 pixels.

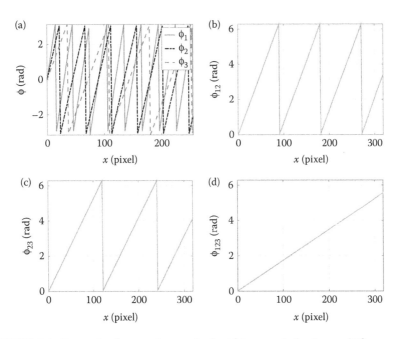

FIGURE 4.1 Example of generating equivalent fringe period using multifrequency phase-shifting algorithm. (a) Cross sections of each frequency phase maps; (b) cross section of ϕ_{12}; (c) cross section of ϕ_{23}; (d) cross section of ϕ_{123}.

Due to very large noise, it is not preferable to directly use the overall equivalent phase $\phi_{eq}(x, y) = \Phi(x, y)$ for 3D reconstruction. Additional step, called *backward phase unwrapping*, is usually required. The backward phase unwrapping essentially starts with the overall equivalent phase $\phi_{eq}(x, y)$ to unwrap the equivalent phase for the second longest fringe period, followed by using unwrapping the phase for the third equivalent phase. For the example given above ($T_1 = 30$, $T_2 = 45$, and $T_3 = 72$ pixels),

$$T_{12} = 90, \tag{4.5}$$

$$T_{23} = 120, \tag{4.6}$$

$$T_{123} = 360. \tag{4.7}$$

To minimize the noise effect, the step-by-step backward phase unwrapping could be

1. Using $\Phi(x, y) = \phi_{123}(x, y)$ to unwrap equivalent phase $\phi_{23}(x, y)$ to obtain unwrapped phase $\Phi_{23}(x, y)$.

2. Using $\Phi_{23}(x, y)$ to unwrap equivalent phase ϕ_{12} to obtain unwrapped phase $\Phi_{12}(x, y)$.

3. Using $\Phi_{12}(x, y)$ to unwrap ϕ_3 to obtain unwrapped phase $\Phi_3(x, y)$.

4. Using $\Phi_3(x, y)$ to unwrap ϕ_2 to obtain unwrapped phase $\Phi_1(x, y)$.

5. Using $\Phi_2(x, y)$ to unwrap ϕ_1 to obtain unwrapped phase $\Phi_1(x, y)$.

Fringe order for each step can be determined through the following equation:

$$k(x, y) = \text{Round} \left[\frac{T^2/T^1 \Phi^2 - \phi^1}{2\pi} \right], \tag{4.8}$$

and the phase can be unwrapped using

$$\Phi^1(x, y) = \phi(x, y) + 2\pi \times k(x, y). \tag{4.9}$$

where Φ^2 is the unwrapped phase for the longer fringe period, ϕ^1 is the phase to be unwrapped, T^2 is the (equivalent) longer fringe period, and T^1 is the (equivalent) shorter fringe period.

The multifrequency phase-shifting method typically requires three frequencies with at least nine fringe patterns. However, for DFP systems,

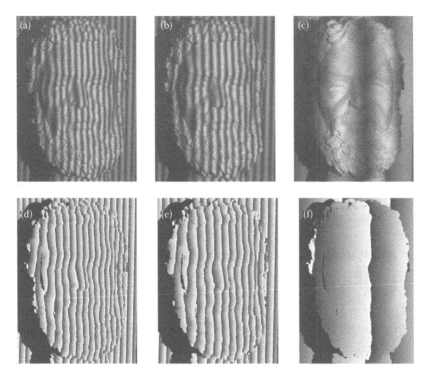

FIGURE 4.2 Example of measuring complex shape 3D geometry using a three-frequency phase-shifting algorithm. (a) One of fringe patterns for T_1 = 30 pixels. (b) One of fringe patterns for T_2 = 36 pixels. (c) One of fringe patterns for T_3 = 231 pixels. (d) Wrapped phase for ϕ_1. (e) Wrapped phase ϕ_2. (f) Wrapped phase ϕ_3.

two-frequency algorithms also work well for special applications [68–70]. Figure 4.2 shows an example of using three frequency phase-shifting algorithm complex 3D imaging. In this example, three fringe periods used were $T_1 = 30$, $T_2 = 36$, and $T_3 = 231$ pixels. The equivalent fringe period $T^{eq} = 815$ pixels. For each fringe period, three phase-shifted fringe patterns with a phase shift of $2\pi/3$ were projected by a DLP projector (Dell MS115HD, Round Rock, Texas) and captured by a CCD camera (The Imaging Source DMK 23U618, Bremen, Germany). The wrapped phase map for each fringe periods was then processed using the multifrequency phase-shifting algorithm.

From those three wrapped phase maps shown in Figure 4.2, the equivalent unwrapped phase can be obtained pixel by pixel using the multifrequency phase-shifting algorithm, as shown in Figure 4.3. Once the unwrapped phase is obtained, 3D shape can be calculated if the system is properly calibrated. In this particular example, a very simple

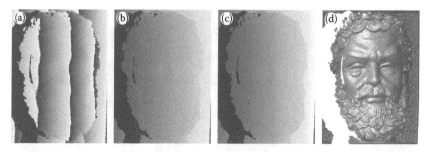

FIGURE 4.3 Example of measuring complex shape 3D geometry using a three-frequency phase-shifting algorithm. (a) Equivalent phase ϕ_{12}. (b) Equivalent phase $\phi_{123} = \Phi(x, y)$. (c) Unwrapped phase Φ_1. (d) 3D reconstructed shape.

reference-plane-based calibration approach was used to convert absolute phase to depth. The reference-plane-based calibration method was detailed in Reference 67.

It is important to note that, for applications where speed is not the priority, this method provides high-quality 3D imaging. However, for high-speed applications, the fewer number fringe patterns used, the higher measurement speeds can be achieved. Therefore, methods have been developed to use fewer patterns for temporal phase unwrapping. Due to the flexibility of generating arbitrary shape patterns, DFP systems enabled some special temporal phase-unwrapping algorithms.

4.3 ADDING MULTIPLE BINARY CODED STRUCTURED PATTERNS

The robustness of determining fringe order, $k(x, y)$, can be substantially improved by using a sequence of binary structured patterns, and this method is usually called hybrid method since it combines phase-shifting method with binary coding method [71]. Essentially, all those binary coding algorithms [17] developed for binary structured light systems can be adopted. These coding methods include simple coding [72], stripe boundary coding [73], gray coding [74], modified coding [75], or any space-time coding method [76]. The difference between different coding methods is the robustness to handle high-contrast object surfaces and the reliability to determine the boundary of the code word changes.

Figure 4.4 illustrates two coding methods that demonstrate the coincidence of the code word changes with the 2π jumps such that the fringe order can be uniquely determined, and thus the phase can be unwrapped

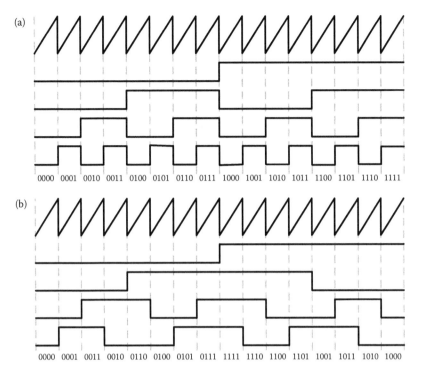

FIGURE 4.4 Example of using binary coded patterns for fringe order determination. (a) Example simple coding method. (b) Example gray coding method.

pixel by pixel (or temporally). There are quite a few types of coding methods, comparing with simple coding, the gray coding is to ensure that at each point there is just one bit change for all the coded patterns and thus the code word—detecting error could be reduced and the unwrapping result can be improved.

Figure 4.5 shows an example of using multiple binary coded structured patterns for absolute phase recovery. Once the absolute phase map is obtained, 3D shape can be recovered. However, due to the artifacts generated by sampling, a computational framework is usually required for final 3D shape reconstruction. The details of removing artifacts are discussed in Section 4.8, and the final reconstructed 3D shape is shown in Figure 4.15 later in the chapter.

This is an extensively employed method that works pretty well for pseudostatic objects. However, when the objects are moving, the motion could cause phase-unwrapping problems even if the gray coding patterns are used. There will be misalignment between the extracted wrapped phase

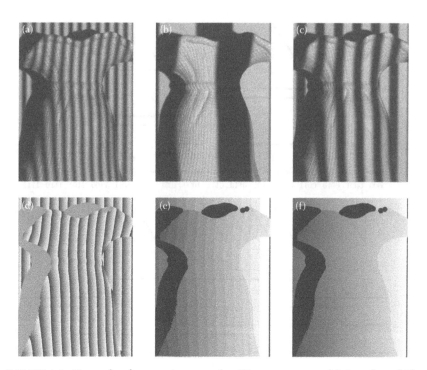

FIGURE 4.5 Example of measuring complex 3D geometry combining phaseshifting with binary coding algorithms. (a) One of the phase-shifted fringe patterns. (b) One of the wide binary patterns. (c) One of the narrow binary patterns. (d) Wrapped phase map. (e) Fringe order from binary patterns. (f) Unwrapped phase map.

and precisely designed code word [74]. Therefore, errors will arise near the jump edges of codeword, and additional computational frameworks are required to properly resolve the associated problems.

4.4 ENCODING FRINGE ORDER INTO PHASE-SHIFTED FRINGE PATTERNS

As aforementioned, the gray coding + phase-shifting technique is one of the major methods by designing a unique code word assigned to each 2π phase-change period to determine fringe order $k(x, y)$. However, the maximum number of available unique code word is limited to 2^M (M is the number of binary patterns used). For example, three binary coded patterns can generate up to eight unique code words. If the code words are embedded into the phase domain through a phase-shifting algorithm [77], more code word could be generated. Specifically, instead of encoding the code word into binary intensity images, the code word is embedded into the phase ranging $[-\pi, +\pi)$ of phase-shifted fringe images (e.g., three images for a three-step phase-shifting algorithm). Since the code word is a finite number that can quantify the phase into discrete levels, this technique has the following merits: (1) less sensitive to surface contrast, ambient light, and camera noises because of the use of phase-shifting method; (2) fast measuring speed since it only requires three additional images to determine fringe order larger than eight (better than a gray coding method).

Figure 4.6 shows the principle of the phase coding framework. Figure 4.6a shows top image one of the phase-shifted fringe patterns to obtain wrapped phase $\phi(x, y)$ with 2π discontinuities. Figure 4.6a bottom image shows one of the phase-shifted fringe patterns with the encoded phase $\phi^s(x, y)$. If the stair changes of a stair phase $\phi^s(x, y)$ is perfectly aligned with the 2π discontinuities, the stairs can be used to determine fringe order for phase unwrapping. If each stair is unique, the stair information can be treated as code word to remove 2π jumps point by point, and thus to obtain the absolute phase. In this research, the code word is represented as the phase ranging from $-\pi$ to $+\pi$ with each stair height as shown in Figure 4.6b. For example, to generate N code words, ϕ^s is quantized into N levels with a stair height of $2\pi/N$, and the phase ϕ^s is encoded with a set of phase-shifted fringe images. By this means, the coding is realized in phase domain. It is well known that the phase is more robust to carry on information than intensity, and is less sensitive to the influence of sensor noise,

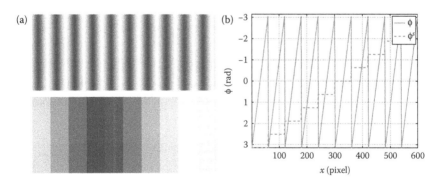

FIGURE 4.6 Illustration of encoding fringe order into phase-shifted fringe patterns. (a) One of the phase-shifted fringe patterns. Top image shows the regular fringe pattern to obtain wrapped phase, and bottom image shows the encoded fringe pattern. (b) Wrapped phase and the code words extracted from encoded fringe patterns. (From Y. Wang and S. Zhang, *Opt. Lett.* 37(11), 2067–2069, 2012. With permission.)

ambient light, surface properties, etc. This proposed technique is inherently better than the gray coding method. Moreover, the maximum unique number it can generate for a gray coding method is 2^M for M binary images. Therefore, it typically requires more than three patterns to recover absolute phase. In contrast, the proposed method only requires three fringe images to generate more than eight unique numbers, as demonstrated by Zheng and Da [78]. Therefore, it has potential to increase measurement speed since fewer images are required to recover one 3D frame.

Here we summarize the procedures that are necessary to retrieve absolute phase.

Step 1: Embed code word into phase with a stair phase function,

$$\phi^s(x, y) = -\pi + \text{Floor}[x/T] \times \frac{2\pi}{N}. \qquad (4.10)$$

Here $\text{Floor}[x/T] = k(x, y)$ generates the truncated integer representing fringe order that removes the decimals of a floating point data while keeps its integer part; T the fringe pitch; the number of pixels per period; and N the total number of fringe periods.

Step 2: Put the stair phase ϕ^s into three phase-shifted fringe patterns,

$$I_1^s(x, y) = I'(x, y) + I''(x, y) \cos(\phi^s - 2\pi/3), \qquad (4.11)$$

$$I_2^s(x, y) = I'(x, y) + I''(x, y) \cos(\phi^s), \qquad (4.12)$$

$$I_3^s(x, y) = I'(x, y) + I''(x, y) \cos(\phi^s + 2\pi/3). \qquad (4.13)$$

Step 3: Obtain the stair phase from the coded fringe patterns through phase wrapping. Since the stair phase is encoded into the value ranging from $-\pi$ to $+\pi$, no spatial phase unwrapping is required.

Step 4: Determine fringe order $k(x, y)$ from the stair phase

$$k(x, y) = \text{Round}[N(\phi^s + \pi)/(2\pi)]. \qquad (4.14)$$

Here $\text{Round}(x)$ determines the closest integer.

Step 5: Convert wrapped phase $\phi(x, y)$ to absolute phase. Once the fringe order $k(x, y)$ is determined, the wrapped phase can be unwrapped temporarily.

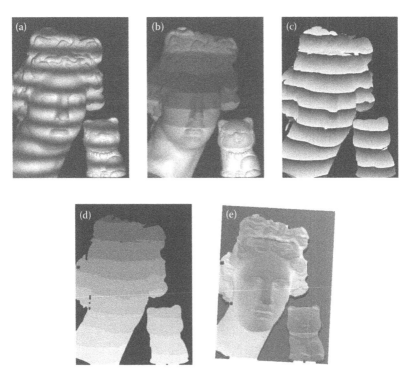

FIGURE 4.7 Example of measuring complex 3D shape using additional phase-encoded patterns. (a) One of the phase-shifted fringe patterns. (b) One of phase-encoded patterns. (c) Wrapped phase. (d) Fringe order recovered from encoded phase patterns. (e) 3D reconstructed shape. (From Y. Wang and S. Zhang, *Opt. Lett.* 37(11), 2067–2069, 2012. With permission.)

Figure 4.7 shows an example of using the phase-encoded patterns for temporal phase unwrapping. Figure 4.7a and b, respectively, shows one of the phase-shifted fringe patterns for phase computation, and one of the phase-shifted fringe patterns for fringe order encoding. From the phase-shifted fringe patterns, the wrapped phase can be obtained, as shown in Figure 4.7c, and from the coded fringe patterns, the fringe order can be resolved, as shown in Figure 4.7d. After unwrap the phase pixel by pixel, 3D shape can be recovered once the system is calibrated. Figure 4.7e shows reconstructed 3D geometry.

Improvements based upon this method have been made by other researchers [78,79]. It is a rather powerful method comparing with intensity-domain binary-coding method for fringe order determination. However, it still requires three additional images for temporal phase unwrapping. If fewer additional images are used, the measurement speed could be improved, and adding one more image instead of three would be an option.

4.5 ADDING A SINGLE STAIR PATTERN

If a stair image is used, and the stair changes are perfectly aligned with the 2π phase discontinuities, the fringe order, $k(x, y)$, can be determined from the stair images [80]. Figure 4.8 illustrates this algorithm. Assume the fringe stripes are vertical, a stair image can be generated as

$$I_s^p(x, y) = \text{Floor}[(x + P/2)/P] \times S. \tag{4.15}$$

Here P is the fringe pitch (number of pixels per fringe period). $\text{Floor}[x]$ generates the truncated integer. S is the intensity level for each stair.

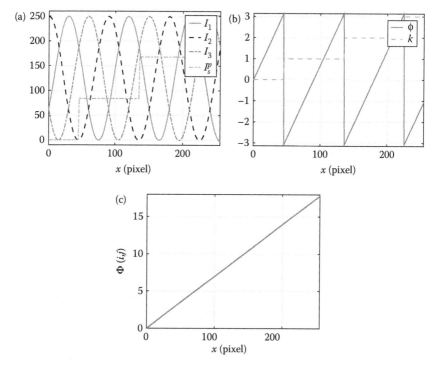

FIGURE 4.8 Schematic diagram of the composite phase-shifting algorithm. (a) Four fringe images used: three phase-shifted sinusoidal fringe images, and one stair image. (b) 2π discontinuities of the phase is precisely aligned with the normalized stair image intensity changes. (c) Unwrapped phase.

Because the object surface reflectivity might not be uniform, normalizing the structured images is necessary to accurately determine integer $k(x, y)$. The normalization procedure is actually straightforward because from Equations 2.21 through 2.23, the maximum and minimum intensity for each pixel can be obtained:

$$I_{min}(x, y) = I'(x, y) - I''(x, y), \tag{4.16}$$

$$I_{max}(x, y) = I'(x, y) + I''(x, y). \tag{4.17}$$

Assume the captured stair image is $I_s(x, y)$, this image can be normalized by the following equation:

$$I_s^n(x, y) = \frac{I_s(x, y) - I_{min}(x, y)}{I_{max}(x, y) - I_{min}(x, y)}. \tag{4.18}$$

Then the integer number $k(x, y)$ can be determined from the normalized stair image as

$$k(x, y) = \text{Round}\left[I_s^n(x, y) \times \frac{R}{S}\right]. \tag{4.19}$$

Here R is the fringe intensity range generated by the computer (e.g., 250) with a maximum value of 255 for a DFP system. Once $k(x, y)$ is determined, the phase can be unwrapped point by point using Equation 4.1.

Theoretically, the integer $k(x, y)$ can be obtained by applying Equation 4.19. For simulations under perfect conditions (e.g., there is no ambient light and the surface reflectivity is uniform), this algorithm works really well [81]. However, it is practically difficult since noise always presents in the real captured image, surface reflectivity may vary from one point to another, and lens defocusing further complicates the problems. Therefore, a complicated computational framework had to be developed to recover absolute phase [80].

Step 1: *Normalize stair image.* This step is to applying Equation 4.18.

Step 2: *Segment wrapped phase into regions through image processing techniques.* This step needs to identify regions so that there is no 2π changes within each region, the unwrapping integer $k(x, y)$ is the same. To increase the robustness of the algorithm, $\pi/4$ instead of π is used as the metric to find the jumps.

The phase edge was detected by a conventional edge-detection algorithm such as Canny filtering. These edge points can be regarded walls to separate different regions. However, because of noises, the wall was not continuous, that is, the edge line segments are broken. To connect those lines, a Wiener filter was applied to the edge image to connect those broken edge line segments. After the edges are correctly located, the regions can be segmented by finding the connected components separated by the edges.

Step 3: Determine $k(x, y)$ for each region for phase unwrapping. Since within each region $k(x, y)$ is a constant, all the intensities of each region on the stair image can be averaged to reduce the noise effect. Applying Equation 4.19 will determine $k(x, y)$ to unwrap the phase.

Step 4: Unwrap edge points. After previous step, the phase should be correctly unwrapped within each region. However, the points on edges have not been processed yet. To process edge points, the phase value can be approximated by interpolating the neighborhood unwrapped phase points $\Phi^0(x, y)$. The true absolute phase value can be determined by

$$\Phi(x, y) = \text{Round}\left[\frac{\Phi^0(x, y) - \phi(x, y)}{2\pi}\right] \times 2\pi + \phi(x, y).$$
(4.20)

Figure 4.9 shows an example of using this approach for 3D shape reconstruction. Figure 4.9a shows one of three phase-shifted fringe patterns, and Figure 4.9b shows the captured stair image. By applying the aforementioned computational framework, the fringe order, $k(x, y)$, can be determined, shown in Figure 4.9c. Temporal phase unwrapping can then be applied to unwrap the wrapped phase pixel by pixel to recover the absolute phase map (Figure 4.9d). Finally, Figure 4.9e shows the reconstructed 3D shape using the calibrated data.

Comparing with the majority temporal phase-unwrapping algorithm, the advantage of this method is obvious: it can achieve higher speed since it only requires one additional pattern for temporal phase unwrapping. The method has been demonstrated its success for relatively complex 3D imaging. However, we found that such a method does not work well for

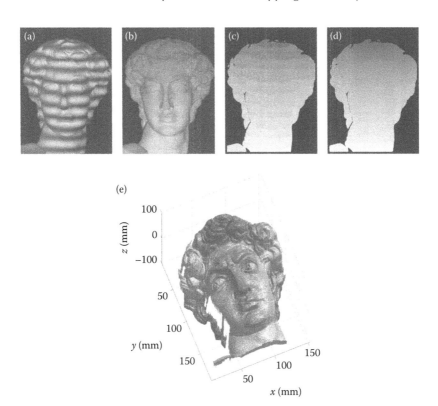

FIGURE 4.9 Example of measuring complex shape 3D geometry using an additional stair image. (a) One of the phase-shifted fringe patterns. (b) Stair image. (c) Fringe order map, $k(x, y)$. (d) Unwrapped phase map. (e) 3D reconstructed shape. (From S. Zhang, *Opt. Laser Eng.* 50(11), 1538–1541, 2012. With permission.)

high-contrast surfaces since the low reflective points have very high noise that could fail the algorithm.

4.6 ADDING A SINGLE STATISTICAL PATTERN

Instead of using many grayscale patterns to represent fringe order, $k(x, y)$, for temporal phase unwrapping, Li et al. [82] proposed a method using a single pattern that only has three grayscale values, 0, 128, and 255, for robust fringe order determinations. These three grayscale values are encoded as 0, 1, and 2. Similar to the stair image, the encoded pattern contains locally unique pseudorandom sequence for each fringe period. Basically, every slit only includes one or two grayscale values; and the slits containing two grayscale values repeat itself vertically. By this means, there are three unique slits with one grayscale values (0, 1, or 2), and three unique slits with

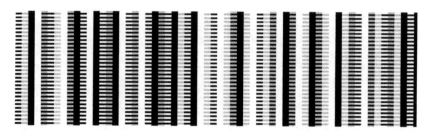

FIGURE 4.10 Example of statistically encoded pattern for temporal phase unwrapping. (Modified from Y. Li et al., *Opt. Express* 18(21), 21,628–21,635, 2010. With permission.)

two grayscale values (0 and 1, 0 and 2, and 1 and 2). These six unique slits are used to generate a pseudorandom sequence for temporal phase unwrapping. The pseudorandom sequence was designed based by searching a Hamilton circuit over a direct graph [83] such that it simultaneously satisfies two properties: (1) any subsequence with a given length (window size) should appear only once in the whole sequence and (2) there are no repeated unique slits in every sequence. Figure 4.10 illustrates one example of the encoded pattern.

The decoding process also involves with image normalization to reduce surface reflectivity influence. Figure 4.11 shows one of successful

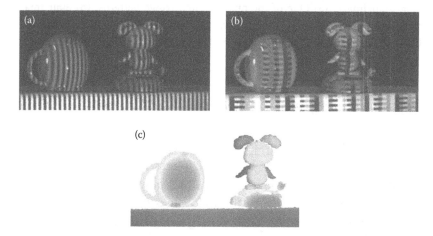

FIGURE 4.11 Example of measuring complex 3D shape using an additional statistically encoded pattern. (a) One of the phase-shifted fringe patterns. (b) Statistically coded pattern. (c) 3D reconstructed shape. (Modified from Y. Li et al., *Opt. Express* 18(21), 21,628–21,635, 2010. With permission; courtesy of Professor Yong Li from Zhejiang Normal University, China.)

measurement results using this approach. Figure 4.11a shows one of the phase-shifted fringe patterns, and Figure 4.11b shows the coded pattern. By using the coded pattern, temporal phase unwrapping can be realized and 3D shape can be reconstructed from the unwrapped phase, as shown in Figure 4.11c.

This method could be less sensitive to noise. However, this method has the limitation of measuring an object smaller than the encoding window; and the decoding process is quite complicated and time consuming, making it difficult for real-time applications. Furthermore, this method may still fail if the object surface has high contrast.

4.7 EMBEDDING STATISTICAL PATTERN INTO PHASE-SHIFTED PATTERNS

All the aforementioned temporal phase-unwrapping methods require at least one more pattern to determine fringe order $k(x, y)$ and thus absolute phase map. Therefore, they all sacrifice measurement speeds. Yet, another approach is to take advantage of the over-constraint system settings by leveraging the passive stereo (camera–camera pair) subsystem and active structured light (camera–projector pairs) subjects. Utilizing two cameras and a projector for 3D imaging is not new and many methods have been developed. In general, the patterns could be random dots [18,20], band-limited random patterns [84], binary-coded patterns [76,85], color structured patterns [86,87], or phase-shifted sinusoidal fringe patterns [88–91]. For these methods, the structured patterns are typically used for stereo matching, rather than directly used for 3D shape reconstruction. Similar to single-camera single-projector systems discussed in Chapter 1, the random-pattern-based methods could achieve high speeds but their spatial resolution is relatively low. The binary-coding methods are relatively robust to noise, but their spatial resolution is limited by both the projector and the camera, albeit is higher than that achieved by random-dots-based methods. Although the use of color could speed up the measurement, the problems associated with color (e.g., color coupling, susceptibility to object surface color) is usually difficult to deal with [92]; and the phase-shifting methods provide the best spatial resolution and great robustness to noise.

Since for high-speed applications, adding more patterns is usually not desirable. Utilizing the geometric constraints and epipolar geometry, the fringe order $k(x, y)$ can be directly determined from the wrapped phase without spatial phase unwrapping [90,91]. Figure 4.12 illustrates the concepts of the trio-geometric constraints of the system. Since no phase

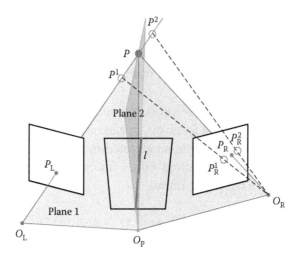

FIGURE 4.12 **(See color insert.)** Epipolar geometry for the trio-sensor system.

unwrapping is performed, any point on the left camera P_L with a phase value of ϕ_0 will corresponds to multiple lines of projector with the same phase value. From epipolar geometry, a 3D point should be on the epipolar plane (plane 1) formed by three points: O_L the optical center of the left camera, O_P the optical center of the projector, and the P_L. Furthermore, optically, the 3D point also lies on the line of $O_L P$. Because of phase constraint, the point must line on the plane (plane 2) formed by the point O_P and phase line l. The intersection point plane 1, plane 2, and line $O_L P$ will give a point P in 3D space. However, because the corresponding phase line on the projector could be multiple if no phase unwrapping is adopted, resulting in multiple 3D point candidates, for example, P^1, P^2, etc. The correct 3D point can be determined by verifying the second camera image. This is because if this 3D point is correct, and the second camera *see* it, the phase value should be close enough. For example, to verify point P^1, the intersection between the right camera sensor and line formed by P^1 and O_R, the optical center of the right camera is a single point, P_R^1. Theoretically, the phase value should also be ϕ_0. Otherwise, P^1 will be rejected.

This method works well. However, the geometric constraints usually require globally backward and forward checking for matching point location, limiting its speed and capability of measuring sharp changing surface geometries. Furthermore, such a system requires accurately calibrate three sensors (two cameras and one projector), which is usually nontrivial.

To overcome the limitations of the method in References 90 and 91, Lohry et al. [93] developed a method that combines the advantages of the stereo approach and the phase-based approach: using a stereo-matching algorithm to obtain the *coarse disparity* map to avoid the global searching and checking associated with the method in Reference 90; and using the local wrapped phase information to further refine the coarse disparity for higher measurement accuracy. Since Lohry's method does not require any geometric constraint imposed by the projector, no projector calibration is required, further simplifying the system development.

The key to the success of the method developed by Lohry et al. [93] is using the stereo algorithm to provide a coarse disparity map. However, none of these parameters, I', I'', or ϕ will provide reliable information about match correspondence for a case like a uniform flat board. To solve this problem without increasing the number of fringe patterns used, this method encoded one or more of these variables to make them locally unique. Since the phase ϕ is most closely related to the 3D measurement quality and an unmodified texture is usually desirable, modifying I'' seems to be the best option.

The encoded pattern was generated using band-limited $1/f$ noise where

$$\frac{1}{20 \text{ pixels}} < f < \frac{1}{5 \text{ pixels}}$$

and with intensity $I_p(x, y)$ such that $0.5 < I_p(x, y) < 1$. Here $I_p(x, y)$ is randomly generated. In Equations 2.21 through 2.23, $I''(x, y)$ was changed to $I_p(x, y)I''(x, y)$. The modified fringe images are described as

$$I_1(x, y) = I' + I_p(x, y)I'' \cos(\phi - 2\pi/3), \tag{4.21}$$

$$I_2(x, y) = I' + I_p(x, y)I'' \cos(\phi), \tag{4.22}$$

$$I_3(x, y) = I' + I_p(x, y)I'' \cos(\phi + 2\pi/3). \tag{4.23}$$

Figure 4.13 illustrates the encoded pattern $I_p(x, y)$ and one of the modified fringe patterns. Since the encoded pattern is still centered around the same average intensity value, the captured texture image or phase should not be affected in theory, albeit the phase SNR may be lower, and the nonlinearity of the projection system may affect texture image quality. Furthermore, any naturally occurring quality map changes caused by object texture or proximity to the projector will be visible from both of the cameras, canceling the effect.

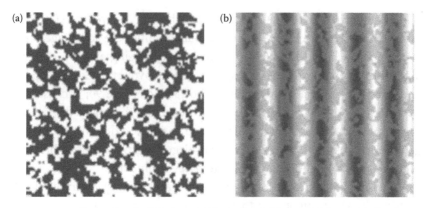

FIGURE 4.13 Example of $1/f$ noise used for encoded pattern. (a) Encoded pattern, $I_p(x, y)$. (b) Modified fringe pattern. (From W. Lohry, V. Chen, and S. Zhang, *Opt. Express* 22(2), 1287–1301, 2014. With permission.)

The 2D varying pattern can improve the cost distinction between a correct match and the other possible disparities. While random pattern stereomatching algorithms have been proposed [94,95], they have been used for the final disparity calculation rather than as an intermediary to matching phase. In this paper, the random pattern is used to match corresponding phase points between two images, without the need for global phase unwrapping. Once the corresponding points have been determined, refinement of the disparity map can proceed using only the wrapped phase locally.

The Efficient LArge-scale Stereo (ELAS) algorithm [16] could be used to obtain an initial coarse disparity map. Since the pattern encoded in data modulation $\gamma(x, y)$ provides great distinctness for many of the pixels, it produces a much more accurate map than just the texture $I'(x, y)$. The encoded random pattern can be converted to an 8-bit grayscale image by scaling the intensity values for quality between 0 and 255 for input into ELAS.

The coarse disparity map provides a rough correspondence between images. However, it must still be refined to obtain a subpixel disparity. While the refinement could be performed on the random pattern itself, refinement using phase has several advantages: the phase is less sensitive to noise and monotonically increases across the image even in the presence of some level of higher-order harmonics.

Unlike the spatial or temporal unwrapping methods that require absolute phase, this method only requires a local unwrapping window along a 3- to 5-pixel line. In a correct match, both the source and the target will lie within π radians, and this constraint can be used to properly align the phases.

The refinement step is defined as finding the subpixel shift τ such that the center of the target phase matches the center of the source phase:

$$x_{\text{target}}(\phi) + \tau = x_{\text{source}}(\phi). \tag{4.24}$$

The relationship between the x coordinate and the phase should locally have the same underlying curve for both the target and the source except for the displacement τ, so $x(\phi)$ can be fitted using a polynomial $a_n \phi^n$, where both the target and the source share the same parameters a_n for $n > 0$.

$$x_{\text{target}}(\phi) = a_0^t + a_1\phi + a_2\phi^2 + a_3\phi^3 \ldots \tag{4.25}$$

$$x_{\text{source}}(\phi) = a_0^s + a_1\phi + a_2\phi^2 + a_3\phi^3 \ldots \tag{4.26}$$

The third-order polynomial fittings was used to refine the disparity. The subpixel shift will be the displacement when $\phi_{\text{source}} = 0$, yielding $\tau = a_0^t - a_0^s$ and a final disparity of $d = d_{\text{coarse}} - \tau$, where d_{coarse} is the coarse disparity for that pixel.

Further study shows that these encoded patterns can be binarized (or dithered) for higher speed 3D imaging [96]. Figure 4.14 shows one example. Figure 4.14a shows the photograph of the system. Figure 4.14b and c, respectively, shows the recovered encoded pattern from the left camera and the right camera. Figure 4.14d shows the wrapped phase from the left camera, and Figure 4.14e shows the wrapped phase from the right camera. These two phase maps are used to remap the pattern and compute the depth map shown in Figure 4.14f. Speckle noise shown in the original depth map can be alleviated by dynamic programming and median filtering. Figure 4.14g and h respectively, shows the result after applying the dynamic programming and subsequently median filtering.

Apparently, this method works fairly well by employing the second camera. Yet, the major limitations are (1) adding the second camera will increase cost of the hardware and the complexity of system design; (2) the area of occlusions increases since three instead of two devices must be able to see a point in order to recover a 3D point; (3) due to the rather slow computational process of phase refinement, it is difficult for this method to achieve real-time processing without dedicated hardware processor (e.g., FPGA or graphics processing unit [GPU]).

Even with such limitations, we believe that adding more camera(s) to 3D imaging system could be the way to go for high-speed applications

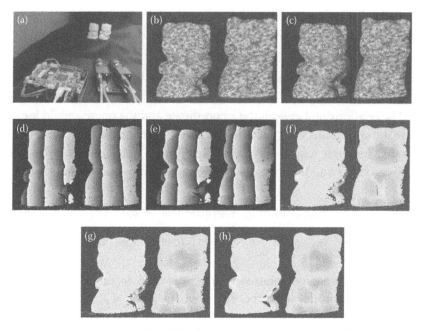

FIGURE 4.14 Experimental results of measuring statues with only three binary patterns. (a) System setup. (b) Coded pattern from left camera. (c) Coded pattern from right camera. (d) Phase map from left camera. (e) Phase map from right camera. (f) Remapped cost aggregation. (g) Result of (f) after applying dynamic programming. (h) Result of (g) after applying median filtering. (From W. Lohry and S. Zhang, *Opt. Express* 22(22), 26,752–26,762, 2014. With permission.)

because the limited frame rates that the projector can refresh patterns and the limited resolutions that a single camera can provide.

4.8 FURTHER DISCUSSION

The temporal phase-unwrapping algorithms presented in this chapter often come with some artifacts that should be properly handled. One of the major problems associated with temporal phase unwrapping is the spiking noise due to sampling of the camera. One of the most effective methods is to apply median filter to locate those incorrectly unwrapped points, and then correct those point phase by adding or subtracting a integer number of 2π. The implementation could be a single pass of a specialized median filter that removes one or two pixels spiking noise due to incorrect phase unwrapping that could be caused by motion or system noise [69]. For example, a median filter with a size of 1×5 that operated in the direction of the phase gradient could be employed to reducing the number of comparisons and branches required. Once the median for a pixel is calculated, the delta between the median phase $\Phi_m(x, y)$ and the original phase $\Phi_o(x, y)$ is taken and rounded after dividing by 2π. This operation results in an integer number $k_0(x, y)$ that can be determined by

$$k_0(x, y) = \text{Round} \left[\frac{\Phi_o(x, y) - \Phi_m(x, y)}{2\pi} \right]. \qquad (4.27)$$

The nonzero $k(x, y)$ indicates a spiking point that can be corrected by subtracting the $k(x, y)$ number of 2π from the original phase to correct the point. That is, the correct phase map can be obtained by

$$\Phi(x, y) = \Phi_o(x, y) - k_0(x, y) \times 2\pi. \qquad (4.28)$$

This filtering stage effectively removes spiking noise yet will not introduce additional artifacts caused by standard median filtering (e.g., smoothing). This is very similar to step 4 of the computational framework discussed in Section 4.5.

Figure 4.15 shows an example of using a median filter to eliminate spiking noise caused by temporal phase unwrapping and further removing the filtering artifacts by applying Equation 4.28. The temporal phase-unwrapping algorithm used in this example is discussed in Section 4.3. Clearly, it shows that directly applying a median filter can eliminate those spiking noise, yet generate artifacts on a surface besides the smoothing effect. After applying the method presented here, filtering artifacts are further eliminated.

One major problem associated with a multifrequency phase-unwrapping algorithm is the noise effect. Even though determined from Equation 4.8 alleviates the noise effect by Round() operator, the noise could still result in incorrect fringe order. This is because the unwrapped phase noise is scaled up by a factor of T^2/T^1. Our study found that one of the most effective methods to reduce this noise influence is to replace $\Phi^2(x, y)$ with the Gaussian smoothed phase $\Phi_g^2(x, y)$; and a large filter size is sometimes needed in order to minimize the problems associated with noise. However, smoothing the phase with a large Gaussian filter could introduce artifacts on the boundaries. To further reduce the boundary problems, one can smooth both $\Phi^2(x, y)$ and $\phi^1(x, y)$ with the same sized filter, and then to find

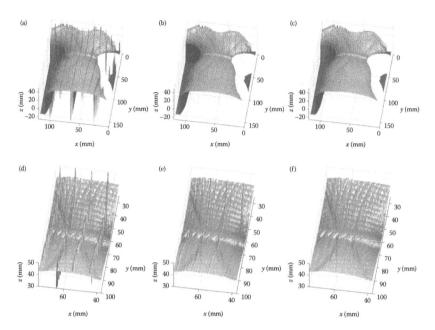

FIGURE 4.15 **(See color insert.)** Example of removing spiking noise caused by temporal phase unwrapping. (a) Raw result without filtering. (b) Result after applying median filtering. (c) Final results after removing filtering artifacts. (d)–(f) Zoom-in views of the above figures.

the fringe order using both smoothed phase maps. It is well known that smoothing phase map in phase domain is equivalent to smooth the sine and cosine functions, and it is usually preferable to smooth the sine can cosine functions instead of phase map directly because existence of 2π discontinuities for a wrapped phase map. Furthermore, to employ a multifrequency phase-shifting algorithm, it is advisable to generate properly fringe patterns such that all ideal patterns generated by computer start from phase 0 to reduce potential problems associated fringe order determination.

4.9 SUMMARY REMARKS

For a DFP system, sine fringe patterns are generated digitally, there are a number of temporal phase-unwrapping algorithms that are not available to conventional interferometry system but have been enabled by the digital technology. This chapter summarizes some of the major temporal phase-unwrapping algorithms, addressed some limitations associated with each of those algorithms. Even though the temporal phase-unwrapping methods are very powerful, spatial phase-unwrapping is still critical for 3D imaging, especially when the measurement speed is one of the major concerns. Next chapter presents a spatial phase-unwrapping algorithm that we developed for real-time 3D imaging system development.

CHAPTER 5

Spatial Phase-Unwrapping Algorithm for Real-Time Applications

A S ADDRESSED IN PREVIOUS CHAPTER, even though temporal phase-shifting algorithm could avoid the spatial phase-unwrapping step, for high-speed applications, a single-frequency phase shifting is more preferable and a spatial phase unwrapping is vital. The large body of literature on phase-unwrapping algorithm developments have been focused on improving the robustness to noise or handling tough situations, and the processing speeds have been a major focus until recently. This chapter mainly presents a special phase-unwrapping algorithm that Zhang et al. [97] developed for real-time applications where the processing speed is a priority, and the major body of this chapter was modified from that publication.

5.1 INTRODUCTION

With the development of the digital technology, real-time 3D imaging is increasingly developed [21]. Even though approaches based on multiple-frequency phase unwrapping have been developed to deliver real-time processing speeds with advanced hardware technology (e.g., GPU) [69,70,98], the single-frequency phase-shifting methods are still extensively adopted due to their faster data acquisition speeds. Therefore, developing a high-speed phase-unwrapping algorithm is still highly desirable for high-speed applications. The single-frequency algorithm has been proven successful for achieving high speeds, for example, Zhang and Huang developed a system that was able to perform real-time 3D shape acquisition, reconstruction, and display at a speed up to 40 frames per second (fps) using a fast three-step phase-shifting algorithm [62]. The phase-unwrapping method used in that system was a scan-line algorithm, which allows real-time performance. However, the simple scan-line phase-unwrapping algorithm encounters significant problems when the phase is not of high quality. Zhang et al. attempted to develop a more robust phase-unwrapping algorithm that will not substantially sacrifice the real-time capability of such a real-time 3D imaging application.

As aforementioned, for a 3D imaging system using a phase-shifting-based method, the phase obtained from fringe images normally ranges from $-\pi$ to $+\pi$. If multiple fringe stripes are used, the phase discontinuities occur every time phase changes by 2π. Phase unwrapping aims to unwrap or integrate the phase along a path counting the 2π discontinuities. The key to reliable phase unwrapping is the ability to accurately detect the 2π jumps. However, for complex geometric surfaces, noisy images, and sharp changing surfaces, spatial phase-unwrapping procedure is usually very difficult [46]. Different phase-unwrapping algorithms have been developed to improve the robustness of the phase-unwrapping process, including the branch-cut algorithms [99–103], the discontinuity minimization algorithm [104], the L^p-norm algorithm [105], the region growing algorithm [106,107], the agglomerative clustering-based approach [108], and the least-squares algorithms [109]. However, these algorithms are generally too slow for high-resolution, real-time 3D shape reconstruction application.

As discussed previously in Chapter 4, multiple-frequency phase-shifting or other temporal phase-unwrapping approaches could achieve high processing speed by using parallel processor (e.g., GPU) because the unwrapping process does not, theoretically, require information from neighboring pixels. However, the temporal phase unwrapping has the following limitations: (1) implementing those algorithms on parallel processors is usually a lot more involved because those algorithms usually require the use of a spatial filter to remove temporal phase-unwrapping artifacts (e.g., spikes) [69]; and (2) they require more fringe patterns images or more hardware (e.g., cameras) with the former slowing down the whole measurement process, and the latter increasing the complexity and cost of the system development. Therefore, spatial phase unwrapping is still extensively adopted in optical metrology, especially for high-speed applications.

Among all spatial phase-unwrapping algorithms developed, the quality-guided phase-unwrapping algorithm usually yields best results [110]. The quality-guided phase-unwrapping algorithms unwrap the phase typically uses a quality map to guide the phase-unwrapping path, and the way of generating quality map and selecting path is vital to the success of phase unwrapping [111]. The unwrapping process starts from the highest quality point and continues to the lower quality ones until it finishes. The quality map is usually constructed based on the first or the second difference between a value and its neighboring pixels. Over the past many years, numerous phase-unwrapping algorithms have been developed including those reasonably fast ones [112–118]. For more references and insights, we recommend the readers refer to the survey paper on quality-guided phase-unwrapping algorithms written by Su and Chen [110], and the survey paper on quality map generation and path selection strategies written by Zhao et al. [111]. It is interesting to note that although some unwrapping error still remains undetected and propagates in a manner that depends on the chosen path, these algorithms are surprisingly robust in practice [46]. However, a conventional quality-guided phase-unwrapping algorithm typically involves a time-consuming point-by-point sorting, making it difficult to be adopted for real-time applications.

Experiments found that it usually took more than 500 ms for a reasonably robust phase unwrapping to unwrap one phase map with an image size of 640 × 480 pixels using an ordinary personal computer (3.2 GHz processor). In contrast, the fast scan-line algorithm can do phase unwrapping rapidly, which has been demonstrated to perform real-time phase unwrapping [62]. If we can reach a compromise between the robust yet slow quality-guided phase-unwrapping algorithm and the less robust yet fast scan-line phase-unwrapping algorithm, the processing speeds of a quality-guided phase-unwrapping algorithm could be substantially improved, and making it possible for real-time applications. Zhang et al. [97] have developed such an algorithm for real-time 3D imaging that is designed for a single connected phase map (i.e., the phase map does not contain isolated areas). The quality map is generated from the gradient of the phase map, then quantized into multilevels. In each level, a scan-line algorithm is applied for unwrapping. Experiments demonstrate that the phase-unwrapping time for three-level algorithm only takes 18.3 ms for one frame and more than 99% real-time acquired facial data with normal facial expressions can be unwrapped correctly. Zhang et al. [97] also implemented this phase-unwrapping algorithm into a real-time 3D imaging system to demonstrate its real-time performance.

Section 5.2 explains the real-time phase-unwrapping algorithm. Section 5.3 shows some experimental validations, and Section 5.4 provides some concluding remarks.

5.2 ALGORITHM

5.2.1 Quality Map Generation

Generating a good quality map is critical for the success of quality-guided phase-unwrapping algorithms, therefore efforts have been devoted to find the proper quality map for robust phase unwrapping, which has been summarized by Su and Chen [110]. This method used two quality maps, the data modulation-based quality map that is available to any phase-shifting algorithm, and the gradient-based quality map that can be derived from the wrapped phase. The former was used to remove the low-quality data points, and the latter was used to guide the path of phase-unwrapping algorithm.

5.2.1.1 Quality Map 1: Background Removal

As discussed in Chapter 2, the data modulation γ in a phase-shifting algorithm provides an indicator for the phase quality with 1 being the best. For a three-step phase-shifting algorithm with equal phase shifts, the data modulation can be determined using Equation 2.25. Previous research found that for a DFP system, the data modulation value can be used to effectively remove the background areas. Therefore, the data modulation was used as the first quality map for background masking.

5.2.1.2 Quality Map 2: Phase-Unwrapping Path Guidance

Once the majority of those low-quality phase points are removed through background masking, the second quality map was used to unwrap those unmasked (or data) points. The quality map used for phase-unwrapping path guiding is a maximum phase gradient map that is defined as

$$Q(i,j) = \max\left\{\Delta_{i,j}^{x}, \Delta_{i,j}^{y}\right\} \quad \text{with } Q(i,j) \in [0,1), \tag{5.1}$$

where the terms $\Delta_{i,j}^{x}$ and $\Delta_{i,j}^{y}$ are the maximum values of the partial derivatives of the phase in x and y directions, respectively.

$$\Delta_{i,j}^{x} = \max\{|\mathbf{W}\{\psi(i,j) - \psi(i,j-1)\}|, |\mathbf{W}\{\psi(i,j+1) - \psi(i,j)\}|\} \tag{5.2}$$

and

$$\Delta_{i,j}^{y} = \max\{|\mathbf{W}\{\psi(i,j) - \psi(i-1,j)\}|, \ |\mathbf{W}\{\psi(i+1,j) - \psi(i,j)\}|\}, \quad (5.3)$$

where $\psi = \phi/(2\pi)$ is the normalized wrapped phase ϕ in Equation 2.24 whose value ranges from 0 to 1. $\mathbf{W}$ is an operator that estimates the true gradient by wrapping the differences of wrapped phase. For example, $\mathbf{W}\{0.3 - (-0.5)\} = -0.2$, $\mathbf{W}\{0.3 - (-0.1)\} = 0.4$. It should be noted that the larger the value of the quality map $Q(i,j)$ in Equation 5.1, the worse the data quality, it is actually a reverse quality map.

5.2.2 Quality Map Quantization

The quality map Q in Equation 5.1 is assumed to be a normal distribution after applying the data modulation masking, with the mean value being $\bar{Q} = \sum_{i=1}^{N}\sum_{j=1}^{M} Q(i,j)/MN$, the standard deviation $\sigma = \sum_{i=1}^{N}\sum_{j=1}^{M}\sqrt{(Q(i,j) - \bar{Q})^2/MN}$. Here the image resolution is $M \times N$. A threshold was first chosen to divide the data into two parts, the points with higher priority to be processed immediately and the points with lower priority to be postponed for later process. In this research, the starting quality value was found to be $th_s = \bar{Q}$. That is, data point set $\{(i,j)|Q(i,j) < th_s\}$ is needed to be processed in the first round. This set forms the first level (level 1). For nth level, the threshold value is $s = th_s + 2^{(n-1)}\sigma$. The last level will unwrap the remaining points.

For the phase data acquired by a real-time 3D imaging system [119], after masking the background with data modulation, more than 80% data points are in level 1. This is because that for a DFP system, the data quality is usually good. Zhang et al. [97] found that the three-level algorithm was sufficient for the system to measure dynamic facial expressions. It should be noted that for the data acquired by the system the more levels used, the slower the processing speed achieved, albeit the more robust the phase-unwrapping algorithm achieved. The extreme case of this algorithm is the conventional quality-guided phase-unwrapping algorithm (will be discussed in Section 5.2.4), in which each time only one pixel with a next lower quality value to proceed.

5.2.3 Scan-Line Algorithm

The scan-line algorithm used in this research is as follows. The process starts from one good point (a point that is not regarded as background and has larger data modulation value larger than 0.7) near the center of the image (x_0, y_0). Then the image is divided into four patches by the horizontal and vertical lines passing the start point, and each patch is unwrapped by a scan-line phase-unwrapping method. Figure 5.1 schematically illustrated the scan-line algorithm. In this scan-line method, one horizontal scan-line scans from the start point to the image border. The scan-line advances vertically from the start point to the image border to scan another row. The neighbors (x_n, y_n) of one scanned point (x, y) can be divided into two groups: the neighbors that faced the start point, namely, the neighbors which have smaller x or y distance to the start point than the scanned point (i.e., $|x_n - x_0| <= |x - x_0|$ and $|y_n - y_0| <= |y - y_0|$); and the neighbors faced the border, which are the neighbors having larger x or y distance to the start point than the scanned point (i.e., $|x_n - x_0| > |x - x_0|$ or $|y_n - y_0| > |y - y_0|$). If at least one of its neighbors faced the start point is unwrapped, a point will be unwrapped and marked as unwrapped. A point with no unwrapped neighbor faced the start point, but has at least one valid neighbor facing the border will be pushed to the stack. Once all points in a patch are scanned, those points in the stack will be popped one by one in reverse order. The popped point with at least one unwrapped neighbor facing the border will be unwrapped, while other points are abandoned. The merit of this method is that each point is only scanned once while it has two chances to be unwrapped. Therefore, it results in better unwrapping results. Since

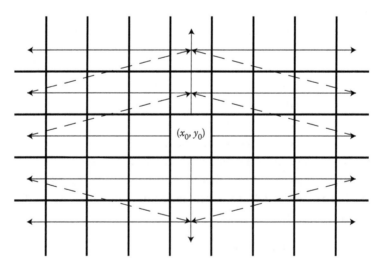

FIGURE 5.1 The schematic diagram of the scan-line phase-unwrapping algorithm. (From S. Zhang, X. Li, and S.-T. Yau, *Appl. Opt.* 46(1), 50–57, 2007. With permission.)

this algorithm is based on line scanning, it even performs faster than the flood-fill algorithm.

5.2.4 Conventional Quality-Guided Phase-Unwrapping Algorithm

The performance of the multilevel quality-guided phase-unwrapping algorithm was compared with the conventional quality-guided phase-unwrapping algorithm that utilizes the phase variance. This section briefly describes this phase-unwrapping algorithm.

For a given phase map, the phase derivative variance is defined as

$$
Z_{m,n} = \frac{\sqrt{\sum(\frac{\partial\psi(i,j)}{x} - \overline{\frac{\partial\psi(i,j)}{x}})^2} + \sqrt{\sum(\frac{\partial\psi(i,j)}{y} - \overline{\frac{\partial\psi(i,j)}{y}})^2}}{k^2}
\tag{5.4}
$$

where for each sum the indexes (i,j) range over the $k \times k$ window centered at the pixel (m, n). $\overline{\frac{\partial\psi(i,j)}{x}}$ and $\overline{\frac{\partial\psi(i,j)}{y}}$ are the averages of these partial derivatives in the $k \times k$ windows. Thus this equation is a rms measure of the variances of the partial derivatives in the x and y directions.

The quality guided path-following algorithm is essentially the flood-fill algorithm, as discussed in Reference 46, in which the order of fill is determined by the quality map. The algorithms operate as follows: A starting pixel with a high-quality value is selected and its four neighbors are examined. These neighbors are unwrapped and stored in a list called the adjoin list. The algorithm then proceeds iteratively as follows: The list pixel with the highest-quality value is removed from the list, and its four neighbors are unwrapped and placed in the list, that is, the pixels are sorted and stored in the list in the order of their quality values. If its neighbor has already been unwrapped, it is not placed in the list. The iterative process of removing the highest-quality pixel from the list, unwrapping its four neighbors and inserting them in the list continues until all the pixels have been unwrapped.

5.3 EXPERIMENTS

To verify the performance of the phase-unwrapping algorithm, the algorithm was used to process fringe images acquired by a 3D imaging system developed by Zhang et al. [119]. Figure 5.2a–c shows three phase-shifted fringe images. From these three images, the phase can be obtained, as whom in Figure 5.2d. Figure 5.2e shows the data modulation, $\gamma(x, y)$, map of these fringe images, from black to white, the value ranges from 0 to 1. Based on the data modulation values obtained, a threshold of 0.25 was

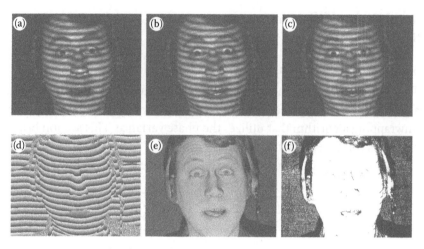

FIGURE 5.2 Example frame for testing multilevel quality-guided phase-unwrapping algorithm. (a)–(c) Three phase-shifted fringe images. (d) Wrapped phase map. (e) Data modulation quality map. (f) Foreground data points (white) after applying a threshold of 0.25 to the data modulation map. (From S. Zhang, X. Li, and S.-T. Yau, *Appl. Opt.* 46(1), 50–57, 2007. With permission.)

applied to remove the background, and a mask map is generated for further processing. Figure 5.2f shows the mask map with 0 being represented as the background.

Figure 5.3 demonstrates the procedures of this phase-unwrapping algorithm. Once the mask map is obtained, shown in Figure 5.2f, one large connected patch is found. The reason for only considering one single connected patch instead of all foreground points is that a real-time system can only perform 3D imaging obtains the absolute coordinates for one single patch. For this connected patch, the quality map is computed using Equation 5.1. The quality map is shown in Figure 5.3b. The histogram of the quality map is shown in Figure 5.3c. The standard deviation of these points is 0.036 and the mean value is 0.039. Hence, the first level threshold is chosen as 0.039. The second level threshold is $0.039 + 0.036 = 0.075$. Since a three-level algorithm is used, the third step unwraps the remaining points.

Figure 5.4 shows the unwrapped points and geometry. The first row shows the unwrapped points after each step which are represented as white. After phase unwrapping, the phase map is obtained, the geometry can then be extracted, as shown in the second row. In this research, the system is calibrated using the method proposed by Zhang and Huang [47]. It should be noted that in each level, the unwrapping algorithm only processes the single

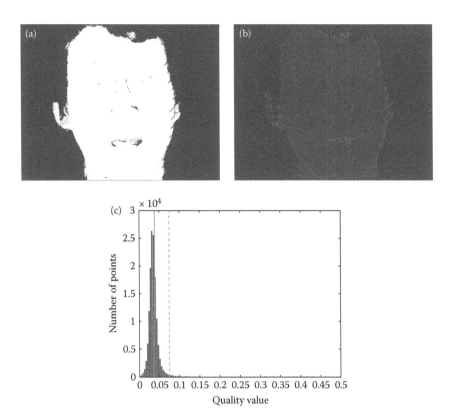

FIGURE 5.3   Quality map and the threshold value for each level. (a) Patch of inter-est, the largest good connected patch. White points are the patch of interest. (b) Quality map. (c) Histograph of the quality map. The solid line shows the mean value which determines the first threshold for the first level. The dashed line shows the threshold used for the second level. (Standard deviation δ: 0.036, mean value: 0.039.) (From S. Zhang, X. Li, and S.-T. Yau, *Appl. Opt.* 46(1), 50–57, 2007. With permission.)

connected patch to boost the efficiency. Those isolated points are postponed to the next step.

Figure 5.5 shows the comparison between the unwrapping results using three different algorithms. It clearly shows that the scan-line algorithm cannot obtain correct geometry. This quality-guided phase-unwrapping algorithm can obtain similar geometry as the traditional variance guided phase-unwrapping algorithm. The difference map is shown in Figure 5.5d. White means unwrapped results are different and black means no differ-ence. There is no difference between the results using the multilevel quality-guided phase-unwrapping algorithm and that using the conventional

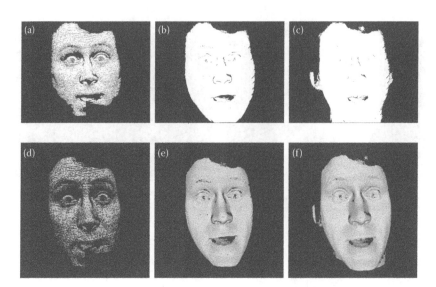

FIGURE 5.4 Unwrapped points after each step. The phase unwrapping starts from the level 1 with the highest quality data points and unwraps these points using the scan-line algorithm, then continues to data points with lower level till finishes. (a) Unwrapping points represented as white after the first level as the first step. (b) Unwrapping points after second step. (c) Unwrapping points when it finishes. (d) Unwrapped geometry after first step. (e) Unwrapped geometry after the second step. (f) Final result. (From S. Zhang, X. Li, and S.-T. Yau, *Appl. Opt.* 46(1), 50–57, 2007. With permission.)

quality-guided phase-unwrapping algorithm. On the contrary, the result obtained by the fast phase-unwrapping algorithm has some area different from that obtained by the conventional variance quality-guided phase-unwrapping algorithm.

An object with a step height was also tested on a flat board. Figure 5.6 shows the results using different algorithms. Again, the fastest algorithm cannot correctly retrieve the geometry, while both this quality-guided phase-unwrapping algorithms can obtain the geometry correctly.

These experiments demonstrated that for the data collected by a real-time structured light system, the quality-guided phase-unwrapping algorithm can achieve comparable results with the traditional variance quality-guided phase-unwrapping algorithm, while the speed of the multilevel quality-guided phase-unwrapping algorithm is much faster. Table 5.1 shows the phase-unwrapping time for a typical face data as shown in Figure 5.2. This table shows that the multilevel quality-guided phase-unwrapping algorithm reduces the processing time of the traditional

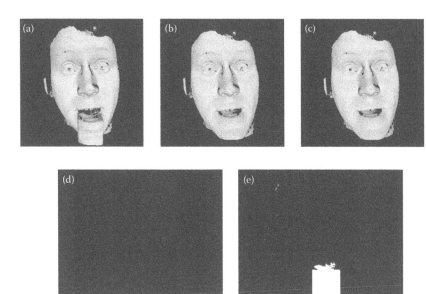

FIGURE 5.5 3D reconstruction results using different phase-unwrapping algorithms. (a) 3D result using fastest scan-line phase-unwrapping algorithm. (b) 3D result using variance quality-guided phase-unwrapping algorithm. (c) 3D result using multilevel quality-guided phase-unwrapping algorithm. (d) The difference map between (b) and (c), black means same and white means different. (e) The difference map between (a) and (b), black means same and white means different. (From S. Zhang, X. Li, and S.-T. Yau, *Appl. Opt.* 46(1), 50–57, 2007. With permission.)

variance quality-guided phase-unwrapping algorithm by approximately 18 times shorter.

To verify the performance of this quality-guided phase-unwrapping algorithm, a number of data sets acquired by a real-time 3D imaging system was also tested. The comparison between algorithms is shown in Table 5.2. Data sets 1–5 are typical human facial expressions. This table shows that for the scan-line algorithm, it can successfully unwrap most of the frames (more 90%). The multilevel quality-guided phase-unwrapping algorithms only fails to a few frames, about 0.1%. A challenging data sequence of 1800 frames, data set #6 in Table 5.2, was also captured to further verify this algorithm. The data set contains the exaggerated and difficult facial expressions. It can be seen that the scan-line algorithm fails more than 86%. The multilevel quality-guided phase-unwrapping algorithm can successfully unwrap about 97%. One may notice that, amazingly, the variance quality-guided

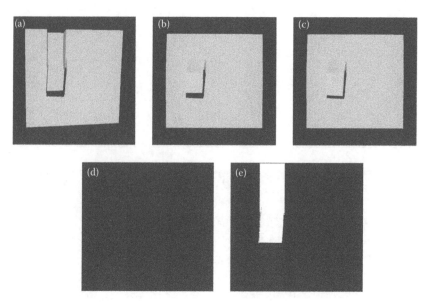

FIGURE 5.6   3D shape reconstruction results for different phase-unwrapping algo-rithms. (a) 3D result using fastest scan-line phase-unwrapping algorithm. (b) 3D result using variance quality-guided phase-unwrapping algorithm. (c) 3D result using multilevel quality-guided phase-unwrapping algorithm. (d) The difference map between (b) and (c), black means same and white means different. (e) The difference map between (a) and (b), black means same and white means differ-ent. (From S. Zhang, X. Li, and S.-T. Yau, *Appl. Opt.* 46(1), 50–57, 2007. With permission.)

phase-unwrapping algorithm can unwrap all frames correctly. These exper-iments demonstrated that this phase-unwrapping algorithm improves the scan-line phase-unwrapping algorithm significantly. It only takes 18.13 ms for a typical face data. Therefore, it is feasible for real-time reconstruction using this algorithm. This phase-unwrapping algorithm was then imple-mented into a real-time system [119]. By employing the fast three-step

TABLE 5.1   Comparison of Phase-Unwrapping Time between Different Algorithms

	Path Following	Trad. Quality-Guided	Multilevel
Unwrapping time (ms)	6.84	505.47	18.13

Note: The computation time was obtained with a Dell Workstation (Pentium 4 3.4 GHz central processing unit, 2 GB Memory), the image resolution is 640 × 480 pixels.

TABLE 5.2   Comparison between Phase-Unwrapping Algorithms

Test Data	Total Frames	Number of Unsuccessful Unwrapping Frames		
		Path Following	Trad. Quality-Guided	Multilevel
Set 1	538	56	0	0
Set 2	538	34	0	1
Set 3	538	12	0	0
Set 4	638	36	0	0
Set 5	538	19	0	2
Set 6	1800	1534	0	71

phase-shifting algorithm [120], real-time 3D shape acquisition, reconstruction, and display at 30 fps were achieved. The achieved speed was similar to the system that Zhang and Huang [62] developed previously with the flood-fill phase-unwrapping algorithm, but with a much better results. Figure 5.7 shows the result during the experiments, the right image is the subject while the image on the left shows the reconstructed geometry in real time.

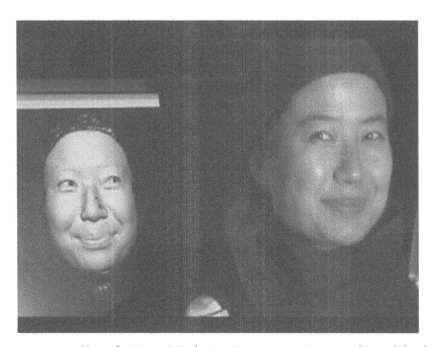

FIGURE 5.7   **(See color insert.)** Real-time 3D reconstruction using this multilevel quality-guided phase-unwrapping algorithm. (From S. Zhang, X. Li, and S.-T. Yau, *Appl. Opt.* 46(1), 50–57, 2007. With permission.)

It should be noted that choosing the threshold is critical to correctly unwrap very complex objects. The thresholds for this three-level algorithm work satisfactorily for the facial data. Experiments found that for data set #6 in Table 5.2, the unsuccessfully unwrapped frames can all be correctly unwrapped by slightly changing the thresholds. A better criteria of setting thresholds for each level might needed to improve the robustness of the algorithm.

5.4 SUMMARY REMARKS

This chapter has presented a special multilevel quality-guided phase-unwrapping algorithm for real-time 3D imaging. It is based on a multilevel quality-guided path-flow method. This algorithm enabled the simultaneously 3D shape acquisition, reconstruction, and display at 30 fps with more than 300,000 points per frame. The purpose of presenting this algorithm is to demonstrate that real-time 3D imaging is feasible even it involves with a robust spatial unwrapping process that is a sequential process in nature. It should not be noted that all the success of this algorithm was based on the testing of human facial expression images, though complex, this algorithm may not work properly for even more complex shape or higher contrast surfaces.

By far, we always assume that high-quality sinusoidal fringe patterns are available. However, for DFP techniques, obtaining high-quality sinusoidal patterns is not straightforward since the projector is usually a nonlinear device. The nonlinear response is often called gamma effect, which should be taken care of before sinusoidal patterns can be generated, which is the focus of next chapter.

CHAPTER 6

Projector Nonlinear Gamma Calibration and Correction

THIS CHAPTER ADDRESSES THE active and passive projector nonlinear gamma compensation methods for phase error reduction. The active method modifies fringe patterns before their projection to ensure sinusoidality; and the passive method, in contrast, compensates for the phase error after capturing those distorted sinusoidal fringe patterns. Our study finds that the active method tends to provide more consistent high-quality fringe patterns regardless the amount of projector's defocusing; yet the effectiveness of a passive method is diminished if measurement condition deviates from calibration condition.

6.1 INTRODUCTION

It is well known that the success of accurate 3D shape measurement using a DFP method heavily relies on the recovered phase quality if a single projector and a single camera is used. This is because such systems recover 3D geometry directly from phase, indicating that any phase noise or distortion will be reflected on final measurements. Among various major error sources, the nonlinear response to input images is one critical error sources to handle if one uses a commercially available digital video projector: this error source often refers to nonlinear gamma effect. Using more fringe patterns [32,121,122] could reduce some high-order harmonics, and thus improve measurement quality. However, using more patterns will sacrifice measurement speeds which is not desirable for high-speed applications. The binary defocusing technology [64,123] could also diminish the nonlinear influence, but it yields lower SNR.

The majority state-of-the-art research focuses on calibrating the nonlinear response of a DFP system and then compensate for the associated error. Though numerous nonlinear gamma calibration and error compensation methods have been developed, they can be broadly classified into two categories: actively modifying the fringe patterns before their projection [124,125] or passively compensating phase error after the fringe patterns are captured [67,126–135]. The mainstream focused on estimating nonlinear gamma coefficients through different algorithms from the captured fringe patterns, and some by directly calibrating the gamma of the projector. Both active and passive methods have been demonstrated successful to substantially reduce the phase error caused by the projector's nonlinear gamma. However, there is no prior study that directly compares the effectiveness of these two types of error compensation methods (i.e., active and passive methods) when the system is not operating under its calibration settings, that is, when the projector has a different amount of defocusing, albeit Reference 125 mentioned the projector's defocusing effect.

This chapter presents a study examining the influence of projector defocusing on the effectiveness of these two different error compensation methods. We found that an active method tends to provide more consistent high-quality fringe patterns regardless the amount of defocusing; yet the effectiveness of a passive method is sensitive to the measurement conditions, albeit the passive method could provide equally good-quality phase under its optimal calibration condition. This research finding coincides with the prior study on binary defocusing technique where the phase error varies with different amounts of defocusing [67], and thus compensating the phase error passively in phase domain is more difficult than actively modifying the fringe patterns before their projection.

Section 6.2 explains two different phase error compensation methods. Section 6.3 shows some simulation results of projector defocusing on sinusoidal and nonsinusoidal patterns. Section 6.4 presents some experimental results. Section 6.5 summarizes this chapter.

6.2 PRINCIPLE

This section presents the passive and active nonlinear gamma calibration methods.

6.2.1 Nonlinear Gamma Model

In the literature, a projector's nonlinear gamma was extensively modeled to be a simple function in the form of

$$I_o = a(I_i)^\Gamma + b, \tag{6.1}$$

where I_o is the output grayscale value for a given input value I_i, a and b are constants, and Γ is the unknown constant to be calibrated. For such a model, estimating the nonlinear effect of the digital video projector essentially is to determine Γ. Constants a and b calibration will not affect the phase quality since they can be optimized by properly adjusting the camera settings. Estimating Γ can be realized through harmonic analysis, least squares, statistical methods, or directly analyzing the phase error by comparing with the ideal phase map.

Over the past decades, we have used over 10 different models as old as Kodak DP900, to Optoma EP739 and PLUS U5-632h, and to the latest models such as LG PB63U and Dell M115HD, the nonlinear gamma varies from one model to another, and even from one projector to another with the same model. The gamma curve of more recent models tends to be smoother than older models with LED projectors being remarkably smooth. Our research found that the nonlinear gamma of the majority projectors we have used in our laboratory does not precisely follow such a simple model if the full range of grayscale values are used. Instead, we found that modeling the projector's nonlinear gamma with a seventh-order polynomial function is sufficient and reliable for all the projectors. That is, the gamma function is described as

$$I_o = c_0 + c_1 I_i + \cdots + c_6 (I_i)^6 + c_7 (I_i)^7, \tag{6.2}$$

where $c_0, c_1, \ldots, c_6, c_7$ are those constants to be calibrated.

6.2.2 Effect of Projector Defocusing

It is well known that the projector defocusing changes the contrast of the projected image, and smooths image sharp features. In theory, defocusing effect is equivalent to applying a low-pass filter (e.g., Gaussian filter) to suppress high-frequency harmonics. It is also clear that the nonlinear gamma effect of the projector converts the ideal input sinusoidal patterns to be nonsinusoidal patterns that contains high-frequency harmonics. In the meantime, those harmonics will be changed by projector defocusing with different amount of defocusing having different influence. Therefore,

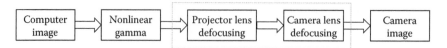

FIGURE 6.1 The whole system response includes both projector's gamma effect and the lens' defocusing.

the equivalent nonlinear gamma effect is actually changed by the amount of defocusing. If the projector is defocused to a certain degree, the nonsinusoidal pattern becomes sinusoidal, making the whole system response to be linear. Here the system response includes the projector's nonlinear gamma and the projector lens defocusing, and the camera lens defocusing, as illustrated in Figure 6.1. It is important to note that we assume that the camera response is linear for all the discussion.

In contrast, if the projector's output pattern is ideal sinusoidal, the projector or camera lens defocusing does not change the sinusoidal nature of the patterns, but rather changes its contrast (or SNR). This is because an ideal sinusoidal pattern has no high-frequency harmonics, and the low-pass filtering will not change its frequency components (with exception that the cut-off frequency is lower than the frequency of the sinusoidal pattern).

The above discussions suggest that if the output image from the projector is ideal sinusoidal, the projector defocusing will not change the structure of the sinusoidal pattern, meaning sinusoidal structure maintains; yet any structures other than sinusoidal will be altered by the effect of defocusing. Therefore, if the DFP system is invariant to the projector defocusing, it requires the output patterns from the projector be ideally sinusoidal, indicating that any passive nonlinear gamma calibration method will not fulfill this requirement; and actively modifying the input fringe patterns is practically the only approach that will work.

6.2.3 Active Error Compensation Method

Active error compensation is quite complex because the calibration condition could be different from case to case; and the modeling should be generic to any sort of captured calibration data. A slightly improved method originally presented in Reference 124 was used in this discussion.

The active method requires the precise gamma response of the projector through calibration. Figure 6.1 indicates that both the projector lens and the camera lens must remain in focus to obtain the projector's nonlinear gamma itself. Therefore, to calibrate the projector gamma, the projector should be focused on the calibration target, and the camera should also be focused on the calibration target.

In practice, the nonlinear response of the whole system can be obtained by projecting a sequence of input grayscale images with different intensity values, I_{ci}, and captured them by the camera. The captured camera image intensities are used as the output data, I_{co}. It should be noted the starting and ending points of the curve are not, respectively, always 0 and 255 to make the approach generic. One should notice that even when both projectors and the cameras are in focus, the calibrated data include the ambient light influence, the nonlinear gamma of the projector, as well as the noise influence [127].

Figure 6.2a illustrates the nonlinear gamma curve and the designed linear response curve. The nonlinear curve can be obtained by projecting a sequence of uniform grayscale images with different grayscale values, I_{ci}, and capturing them by a camera. By analyzing a small area of the camera image, the average value is treated as the output data, I_{co}. It should be noted that the starting and ending points of the curve are not, respectively, always 0 and 255 to make the approach generic.

Since the active calibration method requires modifying the computer generated fringe patterns and predistort the fringe patterns before their projection, the calibration is actually to determine the inverse function of projector's nonlinear gamma. Instead of obtaining polynomial function using Equation 6.2, the inverse function with the output as the x axis was created by fitting. That is, the polynomial function here is actually

$$I_{ci} = a_0 + a_1(I_{co}^s) + \cdots + a_6(I_{co}^s)^6 + a_7(I_{co}^s)^7. \tag{6.3}$$

Here a_k is constant that can be determined by using a set of calibration data.

The objective here is to determine the desired grayscale value, I_d, to be projected for a given value, I_g, such that the projected image will be ideally sinusoidal. Mathematically, I_d can be determined using

$$I_d = a_0 + a_1(I_g^s) + \cdots + a_6(I_g^s)^6 + a_7(I_g^s)^7, \tag{6.4}$$

where

$$I_g^s = \kappa \times (I_g - I_0^{\min}) + I_0^{\min} \tag{6.5}$$

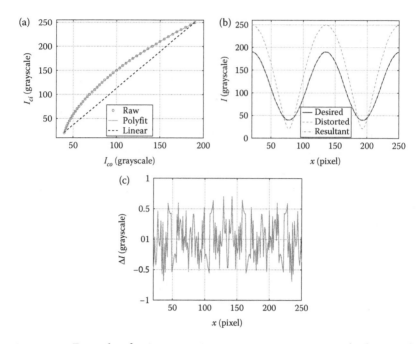

FIGURE 6.2 Example of using an active error compensation method to model nonlinear gamma. (a) Nonlinear gamma curve, fitted polynomial curve, and the desired linear curve. (b) Desired ideal sinusoidal wave, actively distorted wave, and the resultant sinusoidal wave with nonlinear gamma correction. (c) Difference between ideal sinusoidal wave and the resultant sinusoidal wave with nonlinear gamma correction. (From S. Zhang, *Appl. Opt.* 54(13), 3834–3841, 2015. With permission.)

is the modified given input value to consider the fact that the calibrated data range may not be 0–255. Here,

$$\kappa = \frac{I_o^{max} - I_o^{min}}{I_{ci}^{max} - I_{ci}^{min}} \tag{6.6}$$

is the slope of the desired linear response with

$$I_o^{min} = c_0 + c_1[\min(I_{ci})] + \cdots + c_6[\min(I_{ci})]^6 + c_7[\min(I_{ci})]^7, \tag{6.7}$$

$$I_o^{max} = c_0 + c_1[\max(I_{ci})] + \cdots + c_6[\max(I_{ci})]^6 + c_7[\max(I_{ci})]^7. \tag{6.8}$$

where min() and max() are minimum and maximum function, and c_k is from the polynomial function determined using Equation 6.2. Instead of directly using the captured data (i.e., I_{co}^{min} and I_{co}^{max}) as in Reference 124,

I_0^{min} and I_0^{max} are calculated using the fitted polynomial function to reduce noise influence of the raw capture data on both ends.

Figure 6.2b depicts the projected sinusoidal wave, the ideal sinusoidal wave, and the corrected sinusoidal wave using the nonlinear gamma curve shown in Figure 6.2a. Once the distorted curve is modulated by the non-linear gamma function fitted by Equation 6.2, the output curve should be identical to ideal sinusoidal wave. This simulation clearly shows that the projected curve, as expected, perfectly overlaps well with the ideal sinu-soidal wave and the difference is purely random, as illustrated in Figure 6.2c.

6.2.4 Passive Error Compensation Method

The passive error compensation method, in contrast, does not modify the projector's input fringe patterns, but rather determines the phase error from the calibrated gamma curve, and then compensate for the phase error in phase domain. It is straightforward to determine the phase error for each phase value using the following steps if the projector's nonlinear gamma curve is obtained:

- *Step 1:* Compute the ideal phase-shifted fringe patterns. In this case, a three-step phase-shifting algorithm is used as described in Equa-tions 2.21 through 2.23. Only one period of fringe patterns and one cross section of the sinusoidal patterns are only necessary for further analysis.

- *Step 2:* Apply the nonlinear fitted gamma equation as described in Equation 6.2 to generate the distorted curve with gamma effect. Fig-ure 6.3a shows one of the distorted waves by the nonlinear gamma shown in Figure 6.2a.

- *Step 3:* Compute ideal phase, Φ^i, using three ideal sinusoidal waves.

- *Step 4:* Compute the distorted phase, Φ^d, using three distorted waves. Figure 6.3b shows the ideal phase and the distorted phase. It clearly shows that significant phase error is introduced by the nonlinear gamma.

- *Step 5:* Compute phase error by simply taking the difference between ideal phase and distorted phase, that is, $\Delta\Phi(\Phi^d) = \Phi^d - \Phi^i$.

Once the phase error for each distorted phase value is determined, it can be used to compensate for phase error introduced by the nonlinear gamma

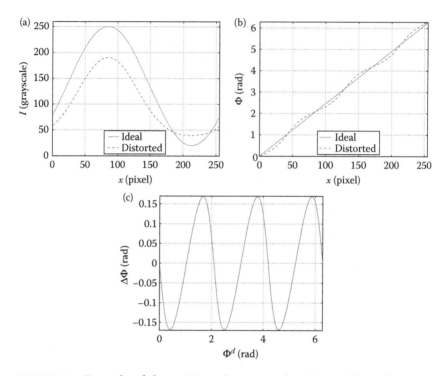

FIGURE 6.3 Example of determining phase error based on calibrated gamma curve. (a) Simulated ideal sinusoidal wave and the distorted wave by gamma effect. (b) Ideal phase Φ^i versus distorted phase Φ^d. (c) Phase error induced by nonlinear gamma. (From S. Zhang, *Appl. Opt.* 54(13), 3834–3841, 2015. With permission.)

effect. Since the error compensation is pixel by pixel for each measurement, the computational cost could be substantial. To reduce computational cost, Zhang and Huang proposed to use a look-up table (LUT) (e.g., 256 elements) [127]. Generating LUT is the process of evenly sampling the error curve and storing the phase error values for each phase value. It is important to note the x axis in Figure 6.3c is *distorted* phase map Φ^d that is the sampling space. The compensation of the phase error can be done by locating the nearest LUT element or involving linear or nonlinear interpolation, and then adding $\Delta\Phi$ to the phase value of that particular point.

6.3 SIMULATIONS

We performed simulations to evaluate the influence of defocusing on sinusoidal and nonsinusoidal fringe patterns. Assume the nonlinear gamma of the projector is shown in Figure 6.4, if ideal sinusoidal patterns are supplied to the projector whose output patterns will be deformed by the nonlinear gamma. As discussed in Section 6.2.1, the nonlinear gamma can be approximated with seventh-order polynomials.

Figure 6.5 shows some simulation results for the influence of defocusing on sinusoidal and nonsinusoidal fringe patterns. Three phase-shifted ideal sinusoidal patterns with a fringe pitch, number of pixels per fringe period, of 45 pixels were generated. These patterns were then modulated by the gamma function using Equation 6.2. Both sinusoidal and nonsinusoidal patterns were smoothed by Gaussian filters with different size to emulate the different amount of defocusing. Gaussian filter size of 1 pixel actually means no smoothing at all. Figure 6.5a shows the nonsinusoidal patterns after applying Gaussian filter size of 1, 15, 31, and 51 pixels; and the cross sections of these patterns are shown in Figure 6.5b. It can been seen that the nonsinusoidal patterns become more and more sinusoidal with the increased size of filters; and their contrast actually decreases. To quantify the sinusoidality of these patterns, the phase error was calculated. The phase error map is determined by taking the difference between the actual phase map, Φ, and the ideal linear phase map, Φ^i, that is, $\Delta\Phi = \Phi - \Phi^i$. All phase maps are obtained by applying the three-step phase-shifting algorithm. Figure 6.5c shows the cross sections of the phase error maps. These results show that applying larger size of Gaussian filter decreases the phase error when the filter size is within a certain range. However, if the filter size is too large, the phase error actually increases. This is because the fringe contrast will be reduced, and the random noise will play important role.

The second row of Figure 6.5 shows the corresponding results when the patterns are ideal sinusoidal. It clearly shows that the phase errors

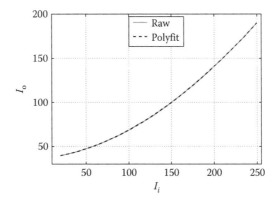

FIGURE 6.4 Nonlinear gamma curve of the projector used for this research.

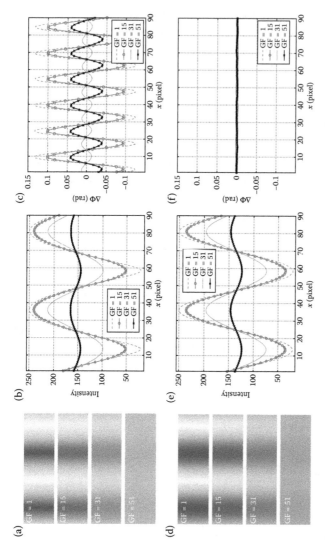

FIGURE 6.5 Simulation results for the influence of defocusing. (a) Nonsinusoidal fringe patterns with different amount of defocusing. GF refers to the Gaussian filter size in pixels. (b) Cross sections of the fringe patterns shown in (a). (c) Cross sections of the phase error maps (rms error 0.0943, 0.0706, 0.0091, and 0.0287 rad for Gaussian filter size of 1, 15, 31, and 51, respectively). (d) Nonsinusoidal fringe patterns with different amount of defocusing. (e) Cross sections of the patterns shown in (d). (f) Cross sections of the phase error maps (rms error 0.0013, 0.0003, 0.0003, and 0.0007 rad for Gaussian filter size of 1, 15, 31, and 51, respectively). (From S. Zhang, *Appl. Opt.* 54(13), 3834–3841, 2015. With permission.)

remain small regardless of the filter size applies, albeit increases when fringe contrast is too low. The simulation confirmed that the defocusing effect substantially change the structures of nonsinusoidal patterns, but have limited influence on ideal sinusoidal patterns.

6.4 EXPERIMENTS

6.4.1 Hardware System Setup

We developed a hardware system to evaluate the performance of these nonlinear gamma calibration approaches. The system includes a DLP projector (Samsung SP-P310MEMX) and a charge-coupled-device (CCD) camera (Jai Pulnix TM-6740CL). The camera is attached with a 16-mm focal length Mega-pixel lens (Computar M1614-MP) with F/1.4 to 16C. The projector resolution and the camera resolution are 800×600 and 640×480, respectively. A uniform flat white board was used as an imaging target for error analysis. It should be noted that the flat board and the camera remain untouched of all the experiments.

The projector's nonlinear gamma curve was obtained by projecting a sequence of unique grayscale images (from 20 to 250) with a grayscale value increment of 5. The camera captures the sequence of images and the grayscale value for each input image is determined by averring a small area (5×5 pixels) in the center of each captured image. Figure 6.4 shows the gamma curve of the projector tested in this research when the projector is in focus.

When ideal sinusoidal patterns are used, the phase error is significant. To demonstrate this, ideal sinusoidal fringe patterns were projected onto the white board and captured three phase-shifted fringe images while the projector is in focus. The phase was calculated by applying a phase wrapping and a temporal phase-unwrapping algorithm. Figure 6.6a shows one cross section of the unwrapped phase map. To better visualize the phase error, the gross slope of the unwrapped phase map was removed.

To quantify phase error, the difference between this phase map and the ideal phase map Φ^i was taken. The ideal phase map was obtained by using the squared binary phase-shifting method [64] with a fringe period of 18 pixels for the projected fringe patterns. The squared binary phase-shifting method can generate high-quality phase without the influence of the nonlinear gamma effect of the projector if a larger number of patterns are used (nine in this case) using the least-squares algorithm [51]. Again, a temporal phase-unwrapping algorithm was used to obtain raw phase that was further smoothed by a large Gaussian filter (e.g., 31×31 pixels) to structural error

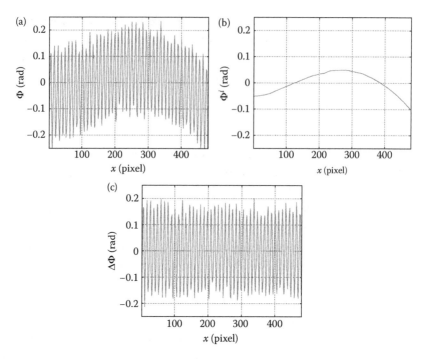

FIGURE 6.6 Phase measurement error of the hardware system without nonlinear gamma correction. (a) Cross section of the unwrapped phase map after removing gross slope. (b) Cross section of the ideal unwrapped phase map after removing gross slope. (c) Cross section of the phase error map without gamma correction (rms of 0.116 mm). (From S. Zhang, *Appl. Opt.* 54(13), 3834–3841, 2015. With permission.)

caused by the system. Figure 6.6b shows the ideal phase after removing its gross slope, which is very smooth, confirming that no obvious systematic structural error was introduced by the ideal phase map, Φ^i. The phase error map was calculated by taking the difference between the capture phase and the ideal phase (i.e., $\Delta\Phi = \Phi - \Phi^i$). Figure 6.6c shows one cross section of the phase error map. If nonlinear gamma is not considered, the phase error is very large with a rms value of 0.116 rad.

6.4.2 Experimental Results for In-Focus Projector

Using the calibrated gamma curve, predistorted the projected fringe patterns using the method discussed in Section 6.2.3 was projected onto the white board. Figure 6.7a shows one cross section of the captured phase after removing its gross slope. Comparing with the result shown in Figure 6.6a, the phase does not have any obvious structural error. Figure 6.7b shows one cross section of the phase error with a rms error of 0.025 rad, proving that the effectiveness of active error compensation.

We then captured three phase-shifted fringe patterns using exactly the same settings except the projector's input fringe patterns are ideal sinusoidal (the same images as those used in Figure 6.7). Figure 6.7c shows the phase error after error compensation using the 512-element LUT discussed in Section 6.2.4. This experiment shows that the passive error compensation can also effectively reduce phase rms error from 0.116 rad to 0.025 rad. Comparing with the active method, the passive method performs equally well.

We also measured a statue to visually compare the differences of these error compensation methods. Figure 6.8 shows the results. Unlike the previous flat board, the statue actually has certain depth variations. As shown

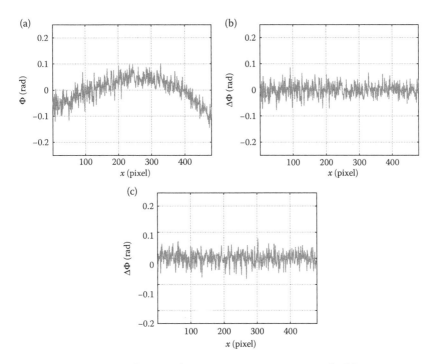

FIGURE 6.7 Passive and active phase error compensation result. (a) Cross section of the unwrapped phase map with active error compensation after removing gross slope. (b) Cross section of the phase error map after active error compensation (rms 0.025 rad). (c) Cross section of the phase error map with passive error compensation (rms 0.025 rad). (From S. Zhang, *Appl. Opt.* 54(13), 3834–3841, 2015. With permission.)

in Figure 6.8b and e, before error compensation, the structural error is very obvious. The active error compensation method provides very high-quality 3D shape measurement without obvious error caused by the nonlinear gamma effect, as shown in Figure 6.8c and f. Figure 6.8d and g shows the results after applying the passive error compensation method. Even though these results are fairly good, the quality is not as high as that of using the active method. We believe this was caused by the fact that the object surface does not always stay in the same amount of defocusing, even when the projector is in focus. These experiments visually demonstrated that the active method outperforms the passive method even when the measurement is close to the calibration condition. It should be noted that all the 3D rendered results were smoothed with a 3 × 3 Gaussian filter to suppress most significant random noise.

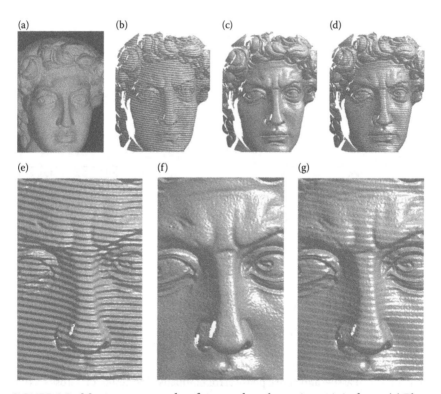

FIGURE 6.8 Measurement results of statue when the projector is in focus. (a) Photograph of the statue. (b) 3D result before gamma compensation. (c) 3D result with active error compensation method. (d) 3D result with passive error compensation method. (e) Zoom-in view of (b). (f) Zoom-in view of (c). (g) Zoom-in view of (d). (Modified from S. Zhang, *Appl. Opt.* 54(13), 3834–3841, 2015. With permission.)

6.4.3 Experimental Results for Out-of-Focus Projector

Since in practical measurement conditions, the object may be placed further away from the gamma calibration plane, meaning the projector may
not be perfectly at the same amount of defocusing. To emulate this effect,
the focal plane of the projector was changed, making the projected image
blurred on the flat board. The same analyses were repeated. Figure 6.9 shows
the results. Comparing results shown in Figures 6.9a and 6.6c, one can
see that the phase error induced by nonlinear gamma is reduced because
of defocusing (rms 0.116 vs. rms 0.080). One may notice that the active
method still performs well (refer to Figure 6.9b), but the passive method has
significant residual structural error (refer to Figure 6.9c). This is because the
defocusing effect actually changed the inherent structures of the fringe patterns if they are not ideally sinusoidal, but does not alter sinusoidal pattern

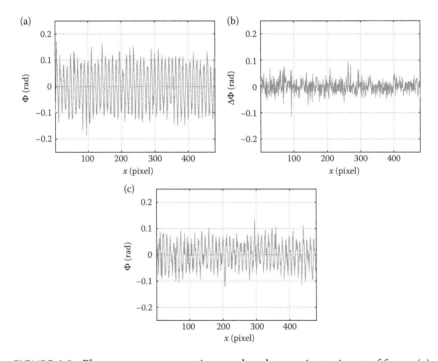

FIGURE 6.9 Phase error compensation results when projector is out of focus. (a)
Cross section of phase error map without any error compensation (rms 0.080 rad).
(b) Cross section of the phase error map with active error compensation method
(rms 0.026 rad). (c) Cross section of the phase error map with passive error compensation method (rms 0.049 rad). (From S. Zhang, *Appl. Opt.* 54(13), 3834–3841,
2015. With permission.)

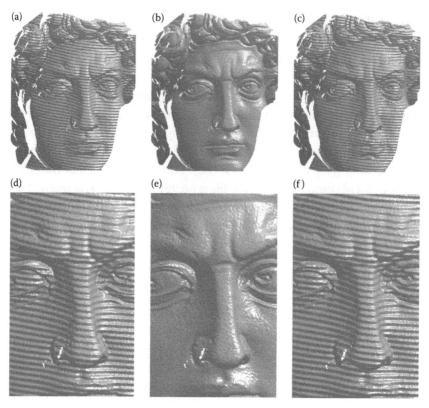

FIGURE 6.10 Measurement results of statue when the projector is out of focus. (a) 3D result before gamma compensation. (b) 3D result with active error compensation method. (c) 3D result with passive error compensation method. (d) Zoom-in view of (a). (e) Zoom-in view of (b). (f) Zoom-in view of (c). (Modified from S. Zhang, *Appl. Opt.* 54(13), 3834–3841, 2015. With permission.)

structures for ideal sinusoidal patterns. One may also notice that the overall phase error for the active method is also slightly increased because of the lower fringe contrast. This coincides with the prior study by Xu et al. [67] demonstrated that the phase error is indeed different for different amount of defocusing, albeit that study shows a different type of nonsinusoidal structured patterns (i.e., squared binary patterns).

Again, the statue was measured when the projector is defocused. Figure 6.10 shows the results. Figure 6.10b and e indicates the active error compensation method still generated good-quality data. However, the passive error compensation method fails to produce high-quality results, as shown in Figure 6.10c and f. These experimental results demonstrated that the active method works much better than the passive method when a different amount of defocusing is used for nonlinear gamma calibration and real measurement. Additionally, comparing Figure 6.10a and d with Figure 6.8b and e, one may also notice that, without applying any error compensation, the measurement results are much better when the projector is out of focus than those when the projector is in focus. This is because the projector defocusing can naturally suppress the nonlinearity of the projector's gamma effect.

Finally, experiments when the projector is at different amounts of defocusing were carried out. Figure 6.11 shows the phase rms error when the projector was at fix different defocusing levels (i.e., from nearly focused, level 1, to substantially defocused, level 6). This figure shows increased defocusing degree (1) diminishes the nonlinear gamma effect of the projector without any compensation; (2) does not obviously affect the active nonlinear gamma calibration method, albeit the phase error increases slightly when the projector is defocused too much; and (3) adversely changes the effectiveness of the passive nonlinear gamma calibration method. It should be noted that this set of data were captured using a newer model projector LG PB63U to show the variations of hardware selection.

We point out the fact that not all system nonlinear gamma curve can be directly calibrated like to the digital video projectors, or the projected patterns can be predistorted. Therefore, for many practical systems, the passive

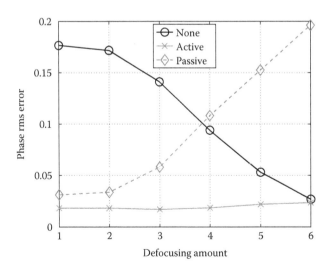

FIGURE 6.11 Comparison of different error compensation methods width different amounts of projector defocusing. (From S. Zhang, *Appl. Opt.* 54(13), 3834–3841, 2015. With permission.)

error compensation methods have to be adopted. Yet, this paper provides the insight that for such systems, if the projection system has different amount of defocusing, a simple error calibration may not be sufficient.

6.5 SUMMARY REMARKS

Experimental data demonstrated that the projector's nonlinear gamma can be calibrated to reduce its influence on the phase quality passively or actively. It reveals that under exactly the calibration condition (e.g., the focus of projector is the same), both methods performed equally well. However, if the projector's focus is changed, the active method does not substantial change, yet, the passive method fails to effectively reduce the phase error caused by the projector's nonlinear gamma curve. Experimental results confirmed that the nonlinear gamma of the projector actually changes with varying amount of defocusing. Therefore, we suggest the use of active gamma calibration approach for 3D shape measurement system development. As such, we provide a sample MATLAB® software source code that reads nonlinear gamma calibration data and generates desired distorted fringe patterns. Two methods were used to generate the predistorted fringe patterns: (1) use the 256-element LUT and (2) directly use the fitted polynomials. The former provides fairly good results, but the latter could produces more accurate patterns since it eliminates the quantization error associated with LUT creation.

CHAPTER 7

Digital Fringe Projection System Calibration

S YSTEM CALIBRATION, WHICH USUALLY INVOLVES complicated and time-consuming procedures, is crucial for any 3D imaging system. This chapter presents two extensively used DFP calibration approaches: the simple reference-plane-based approximation method and the accurate geometric calibration method. The original geometric calibration concept to be presented in this chapter was originally developed by Zhang and Huang [47] and was generalized by Li et al. [137]. The key concept of the geometric calibration method is to enable the projector "capture" images like a camera, thus making the calibration of a projector the same as that of a camera. With this new concept, the calibration of structured light systems becomes essentially the same as the calibration of traditional stereo-vision systems, which is well established. The calibration method is fast, robust, and accurate. This chapter describes the principle of both calibration methods and shows some experimental results to illustrate the calibration ideas.

7.1 INTRODUCTION

It is well known that the calibration accuracy plays a crucial role for high-accuracy 3D imaging for any DFP system. For a DFP system, the calibration essentially is to convert from phase to coordinates in a way to another. Though not highly accurate for large depth range measurement, the simple reference-plane-based methods [138–140] are extensively used mainly because of their simplicity. For a DFP system employing a phase-shifting method, the difference between the measurement phase and reference phase is usually linearly proportional to the depth [67], if the system is properly configured and the measurement depth range is rather small. More accurate calibration requires to calibrate the camera, the projector, and the geometric relationship between these two devices.

Camera calibration has been extensively and rather well studied, and a few very popular methods have been developed. The camera calibration was first performed by using 3D calibration targets [141,142]. By knowing a set of 3D points, the transformation matrix from a 3D space and 2D space can be estimated. Though accurate, these methods require high-precision manufacturing and higher-accuracy measurements of the calibration targets, which are usually very expensive and may not be accessible. To simplify target manufacturing and thus calibration process, Tsai [143] proved that, instead of 3D, precisely moving 2D planar calibration target along the direction perpendicular to the target is sufficient to achieve high calibration accuracy. This method, though accurate and extensive adopted, it typically requires a high-precision linear translation stage, which is, again, quite expensive. To further simply the calibration requirements, Zhang [144] proposed a flexible camera calibration method that does not require precise motion of a 2D planar calibration target, but allow them move at arbitrary poses and orientations, albeit it still requires the knowledge of the target geometry and the preselection of the corner points. Some recent advances of the calibration technique further increase the flexibility by using not-measured or imperfect calibration target [145–148] or by increasing the calibration accuracy by using active calibration targets [149,150].

Yet, the DFP system calibration also requires the projector calibration that is usually not easy and a lot more involved since projector cannot capture images like a camera. Over the years, researchers have developed numerous approaches to calibrate the DFP system. There are attempts to calibrate exact system parameters (i.e., position and orientation) for both camera and projector [151–153]; and these methods could be accurate, but require complicated and time-consuming calibration process. Another general approach is to estimate the relationship between the depth and the phase value through optimization [154–157], which could achieve good accuracy with simplified calibration approach. However, these methods do not take full advantage of the large body of research conducted in stereo-system calibration.

Currently, the popular and extensively adopted DFP system calibration method is to treat the projector as an inverse of an camera; and thus calibrating the projector in a similar manner as calibrating a camera. This is a viable approach, because the camera and the projector are optically the same: have a lens and 2D sensor. Since the camera calibration is well established, the projector calibration would be straightforward. Legarda-Sáenz et al. [158] proposed to use phase as the assistance to establish corresponding points between the projector and the camera and thus can use camera calibrated parameters to calibrate the projector. However, for their method, they used a calibrated camera to calibrate the projector and thus the camera calibration error is coupled into projector calibration. To solve this problem, Zhang and Huang [47] developed a method that makes the projector to *capture* images like a camera. It essentially establishes one-to-one mapping between the projector and the camera with assistance of absolute phase. By this means, the projector calibration is independent of the camera calibration and thus the camera calibration error will not influence the projector calibration accuracy. Following this work, researchers have improved calibration accuracy by using linear interpolation [159], by bundle adjustment [160], or by residual error compensation with planar constraints [161]. All the aforementioned techniques have proven to be successful in calibrating the DFP system. However, they all assume that the projector is at least nearly in focus, which may introduce error if the projector is not.

Another attempt to calibrate the DFP system regardless the projector focus level is a hybrid method, proposed by Merner et al. [162]. In this approach, the depth was calibrated by fitting pixel-wise polynomial fittings through moving a flat surface backward and forward; and the (x, y) coordinate calibration uses a regular camera calibration approach. This method has proven successful to achieve high depth accuracy (e.g., $\pm 50\,\mu m$), but the lateral (along x or y) accuracy is poor (i.e., a few millimeter).

We argue that instead of making the projector *capture* images like a camera, if the one-to-one mapping is established in phase domain, the calibration accuracy will be always ensured [137]. The method essentially is similar to the approach developed by Zhang and Huang [47], but no projector images will be generated. By doing so, the digitization problem associated with generating discrete projector pixels are avoided, and thus higher accuracy could also be achieved with. Since only those feature points (e.g., circle centers or checkerboard corners) are used during the calibration process, the one-to-one mapping is only necessary for those feature points. Moreover, because those feature points are precisely matched, the stereo-system calibration method can be easily adopted.

In this chapter, we mainly present two calibration methods, the simple reference-plane-based method and the projector feature point mapping-based method.

7.2 REFERENCE-PLANE-BASED DFP SYSTEM CALIBRATION

A simple yet extensively employed DFP system calibration method called a reference-plane-based technique that is to convert unwrapped phase to depth z. Figure 7.1 shows the schematic diagram of the DFP system. Points P and I are the center of the exit pupil of the DLP projector and that of the CCD camera, respectively. After the system has been set up, a flat reference plane is measured first whose phase map is used as the reference for subsequent measurements. The height of the object surface is measured relative to this plane. From the point of view of the projector, point D on the object surface has the same phase value as point C on the reference plane, $\Phi_D = \Phi_C^r$. On the camera sensor, point D on the object surface and point A on the reference plane are imaged on the same pixel, $\Phi_A^r \leftarrow \Phi_D$. By subtracting the reference phase map from the object phase map, we obtained

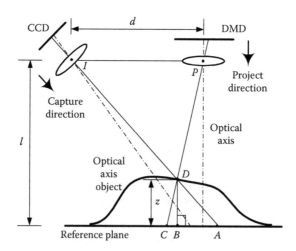

FIGURE 7.1 Schematic diagram of phase-to-height conversion. (From Y. Xu et al., *Appl. Opt.* 50(17), 2572–2581, 2011. With permission.)

the phase difference at this specific pixel:

$$\Delta\Phi_{DA} = \Phi_D - \Phi_A^r = \Phi_C^r - \Phi_A^r = \Delta\Phi_{AC}^r. \tag{7.1}$$

Assume points P and I are designed to be on the same plane with a distance l to the reference plane and have a fixed distance d between them. Assume also that the reference plane is parallel to the plane of the projector lens. Thus, $\triangle PID$ and $\triangle CAD$ are similar, and the height $\overline{DB}$ of point D on the object surface relative to the reference plane can be related to the distance between points A and C

$$\Delta z(x,y) = \overline{DB} = \frac{\overline{AC} \cdot l}{d + \overline{AC}} \approx \frac{l}{d}\overline{AC} \propto \Delta\Phi_{AC}^r = \Phi_D - \Phi_A^r, \tag{7.2}$$

assuming d is much larger than $\overline{AC}$ for real measurement. Combining Equations 7.1 and 7.2, a proportional relationship between the phase map and the surface relative height can be derived point by point. That is, in general,

$$\Delta z \propto \Delta\Phi(x,y) = \Phi(x,y) - \Phi^r(x,y) \tag{7.3}$$

Here $\Phi(x,y)$ is the object phase map and $\Phi^r(x,y)$ is the reference phase map. Assuming the reference plane has a depth of z_0, the depth value for each measured point can be represented as

$$z(x,y) = z_0 + c_0 \times [\Phi(x,y) - \Phi^r(x,y)], \tag{7.4}$$

where c_0 is a constant that can be determined through calibration and z_0 is typically set to be 0.

To determine c_0, a simple calibration can be performed, for example, one can measure a step-height object (top surface plane is parallel to the bottom surface plane) as illustrated in Figure 7.2a with a known step height h (e.g., 10 mm). The phase-shifting method is then applied to measure such an object to obtain unwrapped phase map Φ^o using one of the temporal phase-unwrapping methods. At the same time, the reference plane phase map Φ^r can be obtained, and the difference phase map Φ^d is thus

$$\Delta\Phi = \Phi^o - \Phi^r. \tag{7.5}$$

Plot one cross section of the $\Delta\Phi$ crossing the step block is illustrated in Figure 7.2b. The phase difference between the top surface and the bottom

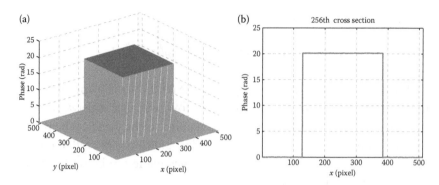

FIGURE 7.2 Determining calibration constant from a step-height object measurement. (a) Phase map. (b) Cross section of phase map.

surface can be obtained using the following steps:

1. Fitting the top surface points with a line $y^t = ax + b_1$

2. Fitting the top surface points with a line $y^b = ax + b_2$

3. Calculating the distance between these two lines $d = |b_2 - b_1|/\sqrt{a^2 + 1}$, here d is in radians

Once the phase difference is obtained, the calibrate constant c_0 can be determined as

$$c_0 = h/d = 10/20 = 0.5 \,(\text{mm/rad}) \tag{7.6}$$

Alternatively, the phase difference can be obtained by fitting the top and bottom surface points as plane functions and find the distance between these two planes.

Determining x and y coordinates are rather easy by scaling with pixel size. Assuming the pixels are eventually spaced, the pixel size can be determined by measuring a known length object.

7.3 ACCURATE DFP SYSTEM CALIBRATION

7.3.1 Camera Pinhole Model

Camera calibration has been extensively studied over the years. A camera is often described by a pinhole model, with intrinsic parameters including focal length, principle point, pixel skew factor, and pixel size; and extrinsic parameters including rotation and translation from world coordinate system to camera coordinate system. Figure 7.3 shows a typical diagram

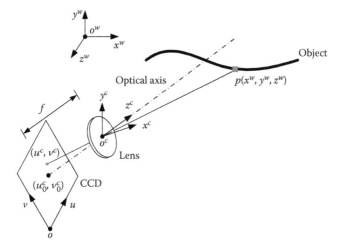

FIGURE 7.3 Pinhole camera model. (From S. Zhang and P. S. Huang, *Opt. Eng.* 45(8), 083601, 2006. With permission.)

of a pinhole camera model, where P is an arbitrary point with coordinates (x^w, y^w, z^w) and (x^c, y^c, z^c) in the world coordinate system $\{o^w; x^w, y^w, z^w\}$ and camera coordinate system $\{o^c; x^c, y^c, z^c\}$, respectively. The coordinate of its projection in the image plane $\{o; u, v\}$ is (u, v). As described in Chapter 1.3, the relationship between a point on the object and its projection on the image sensor can be described as follows based on a projective model:

$$s^c I^c = \mathbf{A}^c [R^c, \quad t^c] X^w, \tag{7.7}$$

where $I^c = [u^c, v^c, 1]^{\mathrm{T}}$ is homogeneous coordinate of the image point in image coordinate system, $X^w = [x^w, y^w, z^w, 1]^{\mathrm{T}}$ homogeneous coordinate of the point in world coordinate system, and s^c a scale factor. $[R^c, t^c]$ are extrinsic parameter matrices, representing rotation and translation between the world coordinate system and camera coordinate system. $\mathbf{A}$ is the camera intrinsic parameters matrix and can be expressed as

$$\mathbf{A}^c = \begin{bmatrix} \alpha^c & \gamma^c & u_0^c \\ 0 & \beta & v_0^c \\ 0 & 0 & 1 \end{bmatrix},$$

where (u_0^c, v_0^c) is the coordinate of principle point, the intersection between the optical axis and the imaging sensor plane, α^c and β^c are focal lengths along u^c and v^c axes of the image plane, and γ^c is the parameter that describes the skewness of two image axes.

Equation 7.7 only represents the linear model of a camera. In practice, the camera lens can have distortion, the nonlinear lens distortion is mainly composed of radial and tangential distortion coefficients, which are usually modeled as a vector of five elements

$$\mathbf{Dist}^c = \begin{bmatrix} k_1^c & k_2^c & p_1^c & p_2^c & k_3^c \end{bmatrix}^{\mathrm{T}}, \tag{7.8}$$

where k_1^c, k_2^c, and k_3^c are the radial distortion coefficients, and p_1^c and p_2^c are tangential distortion coefficients. The radial distortion coefficients can be corrected using the following formula:

$$x' = x(1 + k_1^c r^2 + k_2^c r^4 + k_3^c r^6), \tag{7.9}$$
$$y' = y(1 + k_1^c r^2 + k_2^c r^4 + k_3^c r^6). \tag{7.10}$$

Here, (x, y) and (x', y') refers to camera point coordinate before and after correction, respectively, and $r = \sqrt{x^2 + y^2}$ represents the absolute distance between the camera point and the origin. Similarly, tangential distortion coefficients can be corrected using the following formula:

$$x' = x + [2p_1^c xy + p_2^c(r^2 + 2x^2)], \tag{7.11}$$
$$y' = y + [p_1^c(r^2 + 2y^2) + 2p_2^c xy]. \tag{7.12}$$

7.3.2 Projector Pinhole Model

A projector can be regarded as the inverse of a camera because it projects images instead of capturing them. However, optically, the projector is identical to the camera, each containing one 2D sensor and one lens, and thus modeling a projector is same to that of a camera as

$$s^p I^p = A^p[R^p, \quad t^p]X^w, \tag{7.13}$$

where $I^p = [u^p, v^p, 1]^{\mathrm{T}}$ is homogeneous coordinate of the sensor point in 2D projector sensor coordinate system, and s^p a scale factor. $[R^p, t^p]$ are extrinsic parameters matrix. And the lens distortion is described as

$$\mathbf{Dist}^p = \begin{bmatrix} k_1^p & k_2^p & p_1^p & p_2^p & k_3^p \end{bmatrix}^{\mathrm{T}}, \tag{7.14}$$

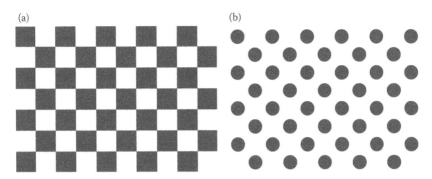

FIGURE 7.4 Typical camera calibration standard target. (a) B/w checkerboard. (b) B/w circle patterns black circles and white background.

7.3.3 Camera Intrinsic Calibration

To calibrate a camera using the flexible camera calibration method developed by Zhang [144], two types of standard targets are mainly used, the black-and-white (b/w) checkerboard, and the b/w circle board, as illustrated in Figure 7.4. After capturing a sequence images of the target with different poses and orientations, the open-source open computer vision (OpenCV) software package, or the MATLAB toolbox developed Bouguet [163], can be used to extract those feature points (e.g., checker corners or circle centers), from which the camera intrinsic parameters and distortion coefficients can be estimated.

7.3.4 Projector Intrinsic Calibration

To accurately calibrate a projector, it is vital to make the projector to "capture" images like a camera, thus making the projector calibration essentially the same as that of a camera, which is well established. Zhang and Huang developed a method using the red/blue checkerboard instead of regular b/w checkerboard for DFP system calibration [47]. This method utilizes the fact that if the checkerboard is illuminated with a uniform white light, the b/w camera does not see the checkerboard and thus serves as a uniform white board; and if the checkerboard is illuminated with a red or blue light, the b/w camera see regular checkerboard. By utilizing this special checkerboard, the projector can capture good quality regular checkerboard images, which will be discussed next.

If a projector can capture images like a camera, its calibration will be as simple as the calibration of a camera. However, the projector obviously cannot directly capture images. Zhang and Huang [47] proposed the concept of using a camera to capture images for the projector by transforming the camera image pixel intensity into projector image pixels. The key to realizing this concept is to establish the precise correspondence between camera pixels and projector pixels, which can be done through phase-shifting techniques. As discussed in Chapter 3, the temporal phase-unwrapping methods give the absolute phase, which does not change from the pattern being projected to the pattern being captured, assuming the noise effect is negligible. If only one directional fringe stripes are used, the correspondence is not unique: the absolute phase value only provides the correspondence up to a phase line (e.g., horizontal line, Φ_h) instead of point. This is a one-to-many mapping: one camera pixel corresponding to many point on the projector. To establish the point correspondence, a second set of fringe patterns that are orthogonal to the first set can be used to obtain another phase line that is perpendicular to the first line, that is, vertical phase line Φ_v. The intersection of these two lines is a point, which is the corresponding point. By this means, the one-to-one mapping between the camera point and the projector point can be established. Figure 7.5 illustrates this framework. In other words, a camera image can be transformed to the projector pixel-by-pixel to form a "virtual" image, which is called the projector image and is regarded as the image "captured" by the projector.

For camera calibration, a standard b/w checkerboard or b/w circle pattern is usually used. However, a b/w checkerboard cannot be used since the fringe images captured by the camera does not have good enough contrast in the areas of the black squares or black circles. To avoid this problem, one can use red/blue checkerboard illustrated in Figure 7.6a could be utilized. Because the responses of the b/w camera to red and blue colors are

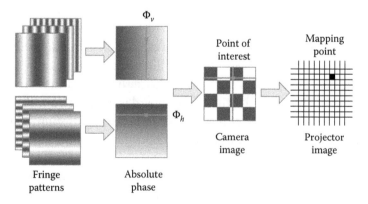

FIGURE 7.5 **(See color insert.)** Projector image generation framework.

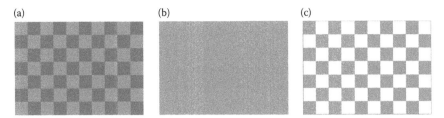

FIGURE 7.6 **(See color insert.)** Effect of illumination on b/w camera image for a red/blue checker board. (a) Red/blue checkerboard. (b) Illuminated with white light and captured by a b/w camera. (c) Illuminated by a red light and captured by a b/w camera. (From S. Zhang and P. S. Huang, *Opt. Eng.* 45(8), 083601, 2006. With permission.)

similar, the b/w camera can only see a uniform board (in the ideal case) if the checkerboard is illuminated by white light as illustrated in Figure 7.6b. When the checkerboard is illuminated by red or blue light, the b/w camera will see a regular checkerboard. Figure 7.6c shows the image of the checkerboard with red light illumination. This checkerboard image captured by the can be mapped onto the projector chip to form its corresponding projector image for projector calibration.

Figure 7.7 shows an example of converting a camera checkerboard image to its corresponding projector image. Figure 7.7a shows the checkerboard image captured by the camera with red light illumination, while Figure 7.7b shows the corresponding projector image. One can verify the accuracy of the projector image by projecting it onto the real checkerboard and checking their alignment. If the alignment is good, it means that the projector image created is accurate.

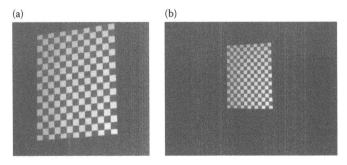

FIGURE 7.7 Example of mapping a camera image to a projector image. (a) Camera image. (b) Corresponding projector image. (From S. Zhang and P. S. Huang, *Opt. Eng.* 45(8), 083601, 2006. With permission.)

This method has been verified successful and achieves great calibration accuracy. However, because the created projector image is a discrete grid that usually involves quantization. This is because the camera pixel may not be precisely aligned with the projector pixel. This could introduce mapping error, though may not be substantial. However, for high-accuracy calibration, this error may not be negligible. Another issue of using checkerboard image is that the phase around the corner may be distorted, as illustrated in Figure 7.8. This figure clearly shows the phase has sharp changes from one checker to another. The phase sharp change could introduce substantial mapping error for corresponding projector pixel determination. This is because the checker corners found on the camera image is not always precise, which could be substantially magnified by this abrupt phase changes. The phase distortion is caused by the different color used for the checkerboard, and the color changes from one to the other around the corner.

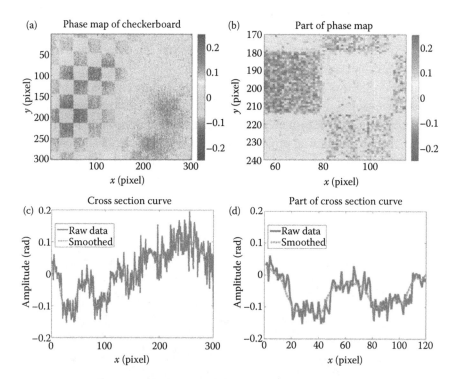

FIGURE 7.8 Phase distortion on the corner points for the checkerboard. (a) Phase map after removing gross slope. (b) Zoom-in view of the phase map. (c) Cross section of the phase map after removing gross slope. (d) Zoom-in view of the cross section.

FIGURE 7.9   White circles on black background.

To circumvent aforementioned problems, Li et al. [137] proposed to use the white circles on black background (as shown in Figure 7.9 instead of red/blue checkerboard for calibration); and further only circle centers are mapped to the projector image without generating the entire projector image since only those feature points are needed for projector calibration. By this means, the error introduced by projector image discretization is avoided. Since white circles are used, around the center of the circle, color does not change, and thus the phase is smooth, as illustrated in Figure 7.10. Under exactly the same condition as those results shown in Figure 7.8, the circle pattern does not have obvious phase distortion around the feature points (i.e., circle centers) used for calibration.

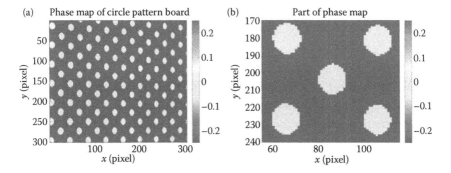

FIGURE 7.10   Phase map is smooth near the center of the circles. (a) Phase map after removing gross slope. (b) Zoom-in view of the phase map.

In summary, the projector calibration requires the following steps:

1. Capture b/w phase-shifted horizontal fringe images projected by the projector with white light illumination, and obtain the horizontal absolute phase map Φ_h.

2. Capture b/w phase-shifted vertical fringe images projected by the projector with white light illumination, and obtain the horizontal absolute phase map Φ_v.

3. Determine the circle centers.

4. Determine projector circle centers for those camera circle centers using one-to-one mapping established through the phase maps.

5. Repeat steps 1–4 a number of times to generate a set of projector circle center points for different poses and orientations.

6. Run the camera calibration software algorithm to calibrate the projector.

7.3.5 Stereo-System Calibration

The purpose of stereo-system calibration is to find the relationships between the camera coordinate system and the world coordinate system and also the projector coordinate system and the same world coordinate system. These relationships can be expressed as

$$X^c = M^c X^w, \tag{7.15}$$

$$X^p = M^p X^w, \tag{7.16}$$

where $M^c = [R^c, t^c]$ is the transformation matrix between the camera coordinate system and the world coordinate system, $M^p = [R^p, t^p]$ is the transformation matrix between the projector coordinate system and the world coordinate system, and $X^c = [x^c, y^c, z^c]^T$, $X^p = [x^p, y^p, z^p]^T$, and $X^w = [x^w, y^w, z^w, 1]^T$ are the coordinate matrices for point p in the camera, projector, and the world coordinate systems, respectively. X^c and X^p can be further transformed to their CCD and DMD image coordinates (u^c, v^c) and (u^p, v^p) by applying the intrinsic matrices A^c and A^p because the intrinsic parameters are already calibrated. That is

$$s^c[u^c, v^c, 1]^T = A^c X^c, \tag{7.17}$$

$$s^p[u^p, v^p, 1]^T = A^p X^p. \tag{7.18}$$

The stereo-system calibration essentially is to determine the extrinsic parameters of the camera and the projector. For this purpose, a unique world coordinate system for the camera and projector has to be established. As discussed in Reference 47, the world coordinate system can established based on one calibration image set with its xy axes on the plane and z axis perpendicular to the plane and pointing toward the system. The extrinsic parameters can be obtained by the same procedures as those for the intrinsic parameters estimation. The only difference is that only one calibration image is used to obtain extrinsic parameters.

However, the extrinsic parameters calibration approach discussed above theoretically works, yet practically does not offer the best calibration accuracy since only one target position is used for extrinsic parameter estimation. The more accurate approach is to utilize all calibration target positions for optimization utilizing the stereo-system calibration algorithm. To do so, instead of estimating the absolute extrinsic parameters for both camera and projector, we can coincide the world coordinate system with the camera or the projector lens coordinate system (see Figure 7.11). For example, if we chose the camera lens coordinate system as the world coordinate system, the extrinsic parameters for the camera is then

$$R^c = \begin{bmatrix} 1 & 0 & 0 \\ 0 & 1 & 0 \\ 0 & 0 & 1 \end{bmatrix}, \tag{7.19}$$

$$t^c = [0, 0, 0]^T. \tag{7.20}$$

Since the world coordinate system is fixed to the camera lens coordinate system, the extrinsic parameter for the projector is essentially the transformation from the projector lens coordinate system to the camera lens coordinate system, which is inherently constant for a given system setup, and thus the extrinsic parameters can be estimated by using more than one target positions for optimization; and higher accuracy will be achieved. Again, by using the mapping circle center pairs, both camera and projector calibration can be simultaneously performed with high accuracy.

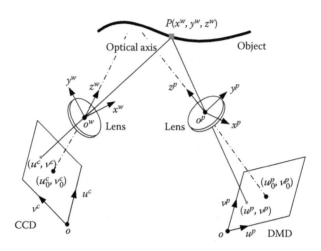

FIGURE 7.11 Pinhole model for stereo system with world coordinate system coinciding with camera lens coordinate system. (From S. Zhang and P. S. Huang, *Opt. Eng.* 45(8), 083601, 2006. With permission.)

7.3.6 Phase-to-Coordinate Conversion

Real-measured object coordinates can be obtained based on the calibrated intrinsic and extrinsic parameters of the camera and the projector. Phase-shifted fringe images can be used to reconstruct the geometry of the surface. In this section, we discuss how to solve for the coordinates based on these four images.

In order to recover 3D geometry, absolute phase $\phi_a(x, y)$ is required. The absolute phase can be obtained either through projecting additional marker on each continuous patch, or using a multifrequency phase-shifting algorithm or a hybrid algorithm to achieve pixel by pixel absolute phase map discussed in Chapter 4. If the absolute phase is obtained, each camera point corresponds one line with the same absolute phase on the projected fringe image [47]. Therefore, we can establish a relationship between the captured fringe image and the projected fringe image through absolute phase,

$$\phi_a(u^c, v^c) = \phi_a^p(u^p) = \phi_a. \tag{7.21}$$

The intrinsic parameter matrices for the camera and the projector are $\mathbf{A}^c$ and $\mathbf{A}^p$, respectively. The extrinsic parameter matrices for a fixed world coordinate system are $\mathbf{M}^c$ and $\mathbf{M}^p$ for the camera and the projector, respectively. Once the system is calibrated, the relationships between the world coordinate system and the camera and projector coordinate systems can be established, we obtain,

$$s^c[u^c, v^c, 1]^T = \mathbf{A}^c\mathbf{M}^c[x^w, y^w, z^w, 1]^T, \tag{7.22}$$

$$s^p[u^p, v^p, 1]^T = \mathbf{A}^p\mathbf{M}^p[x^w, y^w, z^w, 1]^T \tag{7.23}$$

In Equations 7.21 through 7.23, (x^w, y^w, z^w), s^c, s^p u^p, and v^p are unknowns, since there are seven equations, the world coordinate (x^w, y^w, z^w) can be uniquely determined.

Let us assume the camera parameter matrix $\mathbf{H}^c$ and projector matrix $\mathbf{H}^p$ which include both intrinsic and extrinsic parameters,

$$\mathbf{H}^c = \mathbf{A}^c\mathbf{M}^c = \begin{bmatrix} h_{11}^c, & h_{12}^c, & h_{13}^c, & h_{14}^c \\ h_{21}^c, & h_{22}^c, & h_{23}^c, & h_{24}^c \\ h_{31}^c, & h_{32}^c, & h_{33}^c, & h_{34}^c \end{bmatrix}, \tag{7.24}$$

$$\mathbf{H}^p = \mathbf{A}^p\mathbf{M}^p = \begin{bmatrix} h_{11}^p, & h_{12}^p, & h_{13}^p, & h_{14}^p \\ h_{21}^p, & h_{22}^p, & h_{23}^p, & h_{24}^p \\ h_{31}^p, & h_{32}^p, & h_{33}^p, & h_{34}^p \end{bmatrix}. \tag{7.25}$$

Once the absolute phase ϕ_a is obtained, the relationship between the camera coordinates and the projector coordinates can be established, as shown in Equation 7.21.

Since the projector fringe image is composed of uniform stripes, assume the fringe image has a fringe pitch P, the number of pixels per fringe period,

$$u^p = \phi_a^c \times P/(2\pi), \tag{7.26}$$

assuming the absolute phase starts from the edge and increases across the image.

From Equations 7.24 through 7.26, we can obtain

$$\begin{bmatrix} x \\ y \\ z \end{bmatrix} = \begin{bmatrix} h_{11}^c - u^c h_{31}^c & h_{12}^c - u^c h_{32}^c & h_{13}^c - u^c h_{33}^c \\ h_{21}^c - v^c h_{31}^c & h_{22}^c - v^c h_{32}^c & h_{23}^c - v^c h_{33}^c \\ h_{11}^p - u^p h_{31}^p & h_{12}^p - u^p h_{32}^p & h_{13}^p - u^p h_{33}^p \end{bmatrix}^{-1} \begin{bmatrix} u^c h_{34}^c - h_{14}^c \\ v^c h_{34}^c - h_{24}^c \\ u^p h_{34}^p - h_{14}^p \end{bmatrix}. \tag{7.27}$$

It can be seen that the coordinate calculations involve mathematically intensive matrix computations. It will put burden on the central processing unit

(CPU) if all the computations are done by CPU, which makes the real-time reconstruction and display difficult for an ordinary computer. To achieve real-time computation, GPU could be used [164].

7.4 EXAMPLE SYSTEM CALIBRATION

This section demonstrates the calibration process of a DFP system. Figure 7.12 shows a photograph of the system setup. The projector projects phase-shifted fringe images generated by the computer onto the object, then the distorted images will be captured by the camera from another view angle. A synchronization circuit is used to ensure that the camera is triggered with the projector while capturing fringe images. This testing system uses a DLP projector (LightCrafter 3000, Wintech Digital Systems, Carlsbad, California) with a resolution of 608 × 684. It has a micromirror pitch of 7.6 μm. The camera used in this system is a CMOS camera with an image resolution of 1280 × 1024 and a sensor size of 4.8 μm × 4.8 μm (PointGrey FL3-U3-13Y3M-C, Point Grey Imaging, Richmond, Canada). The lens used for the camera is Computar M0814-MP2 lens with a focal length of 8 mm at F/1.4 to F/16.

The system was calibrated using the aforementioned approach. Specifically, the system calibration requires the following major steps:

- *Step 1: Image capture.* The required images to calibrate a system include both fringe images and the actual circle pattern images for each pose of calibration target. The fringe images were captured by projecting a sequence of horizontal and vertical phase-shifted fringe patterns for absolute phase recovery using the multifrequency phase-shifting algorithm. The circle board image was captured by projecting uniform white images onto the board. In total, for each pose, 31

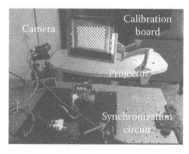

FIGURE 7.12 Photograph of dual-camera structured-light system. (From B. Li, N. Karpinsky, and S. Zhang, *Appl. Opt.* 56(13), 3415–3426, 2014. With permission.)

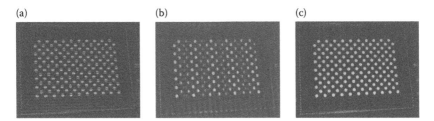

FIGURE 7.13 Example of captured images. (a) Example of one captured fringe images with horizontal pattern projection. (b) Example of one captured fringe images with vertical pattern projection. (c) Example of one captured fringe images with pure white image projection. (From B. Li, N. Karpinsky, and S. Zhang, *Appl. Opt.* 56(13), 3415–3426, 2014. With permission.)

images were recorded for further analysis. Figure 7.13 shows an example of the captured fringe images with horizontal pattern projection, vertical pattern projection, and pure white image projection.

- *Step 2: Camera calibration.* The 18 circle board images were used to find the circle center and then used to estimate the intrinsic parameters and lens distortion parameters of the camera. Both circle center finding and the intrinsic calibration were performed by OpenCV camera calibration toolbox. Figure 7.14a shows one of the circle board image and Figure 7.14b shows the circle centers detected with OpenCV circle center finding software algorithm. The circle-detected circle centers were stored for further analysis.

- *Step 3: Projector circle center determination.* For each calibration pose, the absolute horizontal and vertical gradient phase maps (i.e., ϕ_{ha}^c and

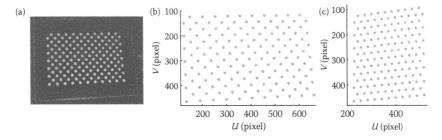

FIGURE 7.14 Example of finding circle centers for the camera and the projector. (a) Example of one calibration pose. (b) Circle centers extracted from (a). (c) Mapped image for the projector from (b). (From B. Li, N. Karpinsky, and S. Zhang, *Appl. Opt.* 56(13), 3415–3426, 2014. With permission.)

ϕ_{va}^c) are obtained using the phase-shifting algorithm. For each circle center, (u^c, v^c), found from step 2 for this pose, the corresponding mapping point on the projector (u^p, v^p) was determined by

$$v^p = \phi_{va}^c(u^c, v^c) \times P/2\pi, \tag{7.28}$$

$$u^p = \phi_{ha}^c(u^c, v^c) \times P/2\pi, \tag{7.29}$$

where P is the fringe period for the narrowest fringe pattern (e.g., 18 pixels in this example).

These equations simply convert phase into projector pixel. The circle center phase values were obtained by bilinear interpolation because of the subpixel circle center detection algorithm for the camera image. Figure 7.14c shows mapped circle centers for the projector. From Equations 7.28 and 7.29, one can deduce that the mapping accuracy is not affected by the accuracy of camera parameters estimation. However, the mapping accuracy could be influenced by the accuracy of circle center extraction and the phase quality. Since the camera circle centers were extracted by standard OpenCV toolbox, one could obtain the coordinates of the circle centers with high accuracy. For high-quality phase generation, in general, the narrower fringe patterns used, the better phase accuracy will be obtained; the more fringe patterns used, the lower the noise effect. Practically, more fringe patterns can be used to reduce the random error effect. In this example, the phase error was reduced by using a nine-step phase-shifting algorithm and the narrow fringe patterns (e.g., fringe period of $T = 18$ pixels). Figure 7.15 shows the reprojection error for both projector and camera, indicating that high calibration accuracy for the projector can be achieved.

- *Step 4: Projector intrinsic calibration.* Once the circle centers for the projector were found from step 3, the same software algorithms for camera calibration were used to estimate the projector's intrinsic parameters. Again, the OpenCV camera calibration toolbox is used in this research. Li et al. [137] found that it was not necessary to consider the lens distortion for the projector, and thus a linear model for the projector calibration was used.

- *Step 5: Extrinsic calibration.* Using OpenCV stereo calibration toolbox and the intrinsic parameters estimated previously, the extrinsic parameters can be estimated. The extrinsic parameter calibrates the

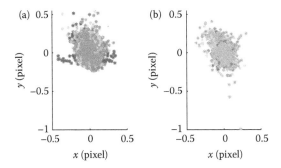

FIGURE 7.15 Reprojection error caused by nonlinear distortion. (a) Error for the camera. (b) Error for the projector. (From B. Li, N. Karpinsky, and S. Zhang, *Appl. Opt.* 56(13), 3415–3426, 2014. With permission.)

transformation from the camera lens coordinate system to the projector lens coordinate system. In other words, the world coordinate system is perfectly aligned with the camera lens coordinates.

In this example, a total of 18 different poses were used for the whole system calibration. The intrinsic parameter matrix for the camera and the projector are, respectively,

$$\mathbf{A}^c = \begin{bmatrix} 1698.02 & 0 & 383.062 \\ 0 & 1691.49 & 294.487 \\ 0 & 0 & 1 \end{bmatrix}, \tag{7.30}$$

and

$$\mathbf{A}^p = \begin{bmatrix} 1019.05 & 0 & 316.763 \\ 0 & 2014.01 & 841.891 \\ 0 & 0 & 1 \end{bmatrix}, \tag{7.31}$$

all in pixels. As aforementioned, though the projector can be accurately calibrated, the camera lens distortion is required, Li et al. [137] found that only the radial distortion k_1 and k_2 in Equation 7.8 needs to be considered. For this particular camera, the lens distortion is

$$\mathbf{Dist}^c = \begin{bmatrix} -0.0905249 & 0.320865 & 0 & 0 & 0 \end{bmatrix}^{\mathrm{T}}. \tag{7.32}$$

Figure 7.15 shows the reprojection error for the camera and projector intrinsic parameter calibration. It clearly shows that the reprojection error is very small (rms 0.15 pixel for the camera and 0.13 pixel for the projector), confirming that both projector and camera can be accurately calibrated. One may notice that there were a few points that have relatively large reprojection errors (around ±0.5 pixels). We believe the large error was caused by the circle center finding uncertainty. As described above, the calibration processes involves reorient and reposition the calibration target to a number of conditions. When the calibration target is parallel to the camera sensor plane, the camera imaging pixels are square and small, and thus circle centers can be accurately determined. However, when the angle between the calibration target plane and the camera sensor plane is larger, the camera imaging pixels are no longer square or small, resulting in difficulty of locating circle centers accurately from the camera image. Nevertheless, the reprojection error is overall very small, all smaller than a pixel size.

The extrinsic parameters for the camera and the projectors are, respectively, in pixels,

$$\mathbf{M}^c = \begin{bmatrix} 1 & 0 & 0 & 0 \\ 0 & 1 & 0 & 0 \\ 0 & 0 & 1 & 0 \end{bmatrix}, \tag{7.33}$$

and

$$\mathbf{M}^p = \begin{bmatrix} 0.952329 & -0.00367422 & 0.305051 & -162.986 \\ 0.0281659 & 0.996716 & -0.0759252 & -146.152 \\ -0.30377 & 0.0808978 & 0.949305 & 95.6518 \end{bmatrix}. \tag{7.34}$$

7.5 CALIBRATION EVALUATION

To verify the performance of the system calibration approach discussed in this chapter, a spherical object was measured, as is shown in Figure 7.16a; Figure 7.16b–e illustrates the three-frequency phase-unwrapping algorithm that were adopted for absolute phase retrieval, which include the phase maps obtained from high-frequency ($T_1 = 18$ pixels), medium frequency ($T_2 = 21$ pixels), and low-frequency ($T_3 = 154$ pixels) fringe patterns, together with the unwrapped phase map after applying the phase-unwrapping algorithm. Then, by applying the absolute phase to coordinate conversion algorithm introduced in Reference 164, one can reconstruct the 3D geometry of the measured object. The measurement results under three defocusing degrees are shown in Figure 7.17. Figure 7.17a–c shows the measurement results, where Figure 7.17a shows the reconstructed 3D

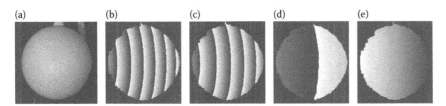

FIGURE 7.16 Absolute phase retrieval using three-frequency phase-unwrapping algorithm. (a) Picture of the spherical object. (b) Wrapped phase map obtained from patterns with fringe period $T = 18$ pixels. (c) Wrapped phase map obtained from patterns with fringe period $T = 21$ pixels. (d) Wrapped phase map obtained from patterns with fringe period $T = 154$ pixels. (e) Unwrapped phase map by applying the temporal phase-unwrapping algorithm with three frequencies. (From B. Li, N. Karpinsky, and S. Zhang, *Appl. Opt.* 56(13), 3415–3426, 2014. With permission.)

surface. The smooth spheric surface indicates a good accuracy. To further evaluate its accuracy, a cross section of the sphere was evaluated by fitting it with an ideal circle. Figure 7.17b shows the overlay of the ideal circle and the measured data points. The difference between these two curves is shown in Figure 7.17c. The error is quite small with a rms error of 0.071 mm or 71 μm. It is important to note that the whole volume of the calibration board poses was around $150(H) \times 250(W) \times 200(D)$ mm^3. These experimental results clearly demonstrate that for such a large calibration volume, the phase-domain-based mapping method can achieve pretty high accuracy.

To further evaluate the calibration accuracy, the lengths of two diagonals on calibration board under the aforementioned three different defocusing

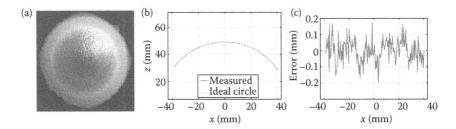

FIGURE 7.17 Measurement result of a spherical surface, the rms errors estimated is ±71 μm. (a) Reconstructed 3D result. (b) A cross section of the 3D result and the ideal circle. (c) Measurement error comparing with the ideal circle. (From B. Li, N. Karpinsky, and S. Zhang, *Appl. Opt.* 56(13), 3415–3426, 2014. With permission.)

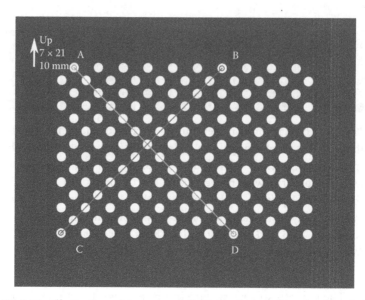

FIGURE 7.18 Illustration of the measured diagonals on calibration board. (From B. Li, N. Karpinsky, and S. Zhang, *Appl. Opt.* 56(13), 3415–3426, 2014. With permission.)

degrees were measured and compared against heir actual lengths measured using a highly accurate digital caliper. The two measured diagonals $\overline{AD}$ and $\overline{BC}$ are shown in Figure 7.18, where $\overline{AD}$ is formed by top left and bottom right circle center pixels, and $\overline{BC}$ is formed by the rest two circle center pixels. It is worth to note that circle centers were detected automatically with subpixel accuracy through Hough transform, and the 3D coordinate of the subpixel were obtained through bilinear interpolation. The measurement results are shown in Table 7.1. It again illustrates that a good measurement accuracy can be achieved. The measurement error is around 0.20 mm. Considering the lengths of the diagonals (around 183.10 mm), the relative error is quite small (∼0.12%). The major sources of error could be the error introduced by circle center detection and bilinear interpolation of 3D coordinates. Moreover, the accuracy is also subject to the precision of caliper measurement.

TABLE 7.1   Measurement Result of Two Diagonals on Calibration Board

System Setup	$\overline{AD}$ (mm)	Error (mm)	$\overline{BC}$ (mm)	Error (mm)
	182.90	0.20	183.50	0.36
Actual	183.10	NA	183.14	NA

Finally, further study by Li and Zhang [165] has demonstrated that by using patterns with the optimal fringe angles [166], more consistent high accuracy could be achieved. Readers are encouraged to use the optimal angle idea to calibrate a structured light system.

7.6 SUMMARY REMARKS

Even though there are numerous calibration approaches developed over the years, the difference among those methods is usually between achievable accuracy and calibration complexity. For the DFP system, both reference-plane-based method and geometric calibration method are extensively adopted. The former is very simple, and can achieve good calibration accuracy within a small range if the system is properly configured. However, overall, the calibration accuracy is not high for large scale measurements. The latter becomes more popular recently because open-source calibration software packages can be directly implemented to achieve great accuracy. The calibration is probably the most important step of high-accuracy 3D imaging system development; and yet probably one of the most difficult steps. In the coming chapters, we demonstrate how to develop a practical 3D imaging system, and achieve higher speeds.

CHAPTER 8

Hands-On Example of System Design and Development

T HIS CHAPTER DETAILS THE design, optimization, calibration, and test of a DFP system. The objective of this chapter is to demonstrate how to develop a 3D imaging system based on a DFP technique through an example specifications and available camera and projector.

8.1 PROBLEM STATEMENT

The objective of this chapter is to develop a compact 3D imaging system based on DFP techniques to achieve sensing area of larger than 250 mm on both horizontal and vertical direction using the following hardware components:

- Camera: The imaging source DMK 23U618

- Projector: Dell M115HD DLP projector

- Signal generator: Arduino A000066 development boards and kits

Based on our prior experience, the angle between projector and camera optical axis should be between 12° and 15° to achieve good sensitivity and compact design; and the shortest focal length lens we use without substantial lens distortion is 8 mm.

(a)   (b)   (c)

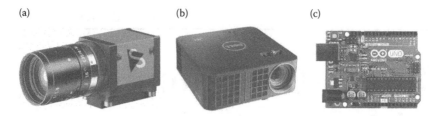

FIGURE 8.1 Hardware components used for system design. All images are obtained from the corresponding manufacturer's Web site. (a) CCD camera; (b) DLP projector; (c) Arduino circuit board.

Figure 8.1 shows the images of the hardware components used.

8.2 MECHANICAL DESIGN AND INTEGRATION

8.2.1 Detailed Specifications for Hardware Components

The critical specifications for the camera that would be valuable for reference:

- Resolution: 640×480

- Maximum frame rate: 120 fps

- Sensor format: $1/4''$ CCD

- Pixel size: $5.6\,\mu m \times 5.6\,\mu m$

- Data connection type: USB 3.0

- Shutter: Global

- Lens mount: C/CS

- External trigger: Yes

- External trigger interface: Hirose male connector pin #11 for trigger in positive and pin #12 for trigger in negative

The critical specifications for the projector that would be valuable for reference:

- Brightness: 450 ANS Lumens (max)

- Native resolution: WXGA 1280×800

- Maximum frame rate: 120 fps

- Projector lens: F-stop: F/2.0; Focal length: $f = 14.95$ mm, fixed lens

- Projection distance: $3.18'$–$8.48'$

- Sensor format: 0.45″ S450 DMD DarkChip3™

- Pixel size: $7.6\,\mu m \times 7.6\,\mu m$

- Video input: HDMI 480 i/p, 576 i/p, 720p, and 1080p

- Light source: LED

The critical specifications for the signal generator are

- Number of bits: 8 bits

- Clock speed: 16 MHz

8.2.2 Lens Selection

The design process starts with selecting a lens that covers as much projection area as possible, and thus aligning the short imaging direction of the camera with the projector is a natural choice. From the camera specifications, we can compute the CCD physical size as

$$c^w = 640 \times 5.6\,\mu m = 3.584\,mm, \tag{8.1}$$

$$c^h = 480 \times 5.6\,\mu m = 2.588\,mm. \tag{8.2}$$

Therefore, the optimized design will use the height of the CCD sensor as the optimization parameter.

The lens selection starts with finding its focal length, ideally, the camera imaging area covers the whole projection area, namely, the imaging height s^h is the same as the projection screen height h. As specified, the minimum screen height is 250 mm, or ideally, the lens will have $s^h = 250$ mm.

The next question is to find out the distance d such that the projection screen height is 250 mm. Figure 8.2 illustrates the geometric dimensions of the projection screen. The key here is to find the half projection angle horizontally θ and vertically α. As the projector use off-axis projection, its vertical projection is asymmetric and its horizontal projection is almost symmetric. If the offset ratio is 100%, meaning that the lowest projection horizontal line does not have any angle from the bottom plane, the projection screen size and the distance d can be found on Table 8.1. From which,

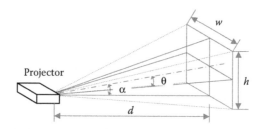

FIGURE 8.2 Lens imaging system for the camera.

TABLE 8.1 Example Projection Screen with an Offset Ratio of 100%

Desired Distance, d, (m)	Screen Size w (cm) × h (cm)	Height (cm)
97	65 × 40	40
129	86 × 54	54
161	107 × 67	67
194	129 × 81	81

we can calculate

$$\alpha = \tan^{-1}\left[\frac{h/2}{d}\right] \approx 11.7°, \tag{8.3}$$

$$\theta \approx \tan^{-1}\left[\frac{w/2}{d}\right] \approx 18.5°. \tag{8.4}$$

Based on the determined angle, α, we can determine the distance L to achieve screen size of 250 mm in the smaller direction, that is,

$$L = \frac{250/2}{\tan \alpha} \approx 606 \, \text{mm}. \tag{8.5}$$

The lens geometry, as illustrated in Figure 8.3, tells that

$$\frac{c^h}{s^h} = \frac{f}{L-f}. \tag{8.6}$$

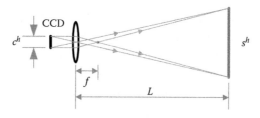

FIGURE 8.3 Lens imaging system for the camera.

Here $c^h = 480 \times 5.6\,\mu m = 2.688\,mm$ is the physical dimension of the camera sensor and $s^h = 250\,mm$ is the desired sensing dimension.

Given $L = 606\,mm$, we can determine the focal length for camera lens,

$$f = \frac{Lc^h}{c^h + s^h} \approx 6.45\,mm. \qquad (8.7)$$

The focal length (f) determined from maximizing projection screen is shorter than 8 mm, which is the shortest we prefer using in our practical measurement system. Therefore, instead of maximizing the projection area, we will use the shortest focal length $f = 8\,mm$ for design optimization.

To have a projection screen size of 250 mm in the shorter dimension with the 8 mm focal length lens, we can compute the distance using L by reorganizing Equation 8.6 as

$$L = \frac{s^h + c^h}{c^h} \times f \approx 752\,mm. \qquad (8.8)$$

8.2.3 Geometric System Design

Once the lens is selected, the next step is to design the geometric relationship between projector and camera to satisfy the specifications. We have already computed $\alpha \approx 11.7°$, and the camera should be rotated toward the projector by at least 0.3°. To be safe, we chose 1.3°, and making the angle between projector and camera optical axis to form an angle of

$$\beta = 13°. \qquad (8.9)$$

Referring to Figure 8.4, we can compute distance between projector and camera

$$B \approx 172.8\,mm. \qquad (8.10)$$

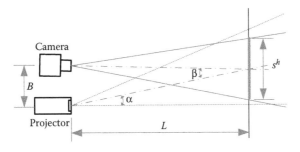

FIGURE 8.4 Lens imaging system for the camera.

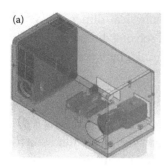
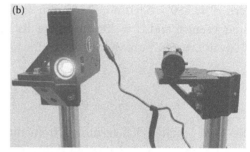

FIGURE 8.5 **(See color insert.)** 3D system design and testing. (a) The whole system design based on the geometric constraint. (b) Testing system before sending the design for fabrication.

Once the geometric constraints have been figured out, the next step is to use a design software (e.g., Solidworks) to design the integrated system to ensure everything looks properly configured. Figure 8.5a shows our design using Solidworks. From this design, we can clearly see that there are not any conflicts between parts so the design can be sent to a machine shop for fabrication.

It is important to note that for this particular design, no additional fans are used for extra cooling, although a fan might be necessary for the most DFP systems. One needs to pay extra attention to cooling, if one is not sure about it, using an extra cooling fan is always a good idea since heating could cause a lot of problems. For example, if the projector temperature is too high, the focal length of the projector lens could change, failing the precalibrated system.

It is also important to note that, before sending the designed parts for fabrication, it is always advisable to roughly set up a system and do some quick testing to ensure that everything works as designed. For example, Figure 8.5b shows a flexible and simple setup that could be developed for testing. One can print out the region with desired field of view (FOV), stick the paper on a flat surface, put the flat surface near the designed distance L from the system, and roughly align the center of the FOV with projection illumination, as shown in Figure 8.6a. Take an image with the camera attached with a designed focus length lens, if the camera can capture the full FOV, as illustrated in Figure 8.6b, the design should be good for fabrication.

8.3 CALIBRATION TARGET

The calibration target size is essentially determined by the designed FOV for the system, in this particular case, it is approximately $250 \times 333 \, \text{mm}^2$.

(a)   (b)

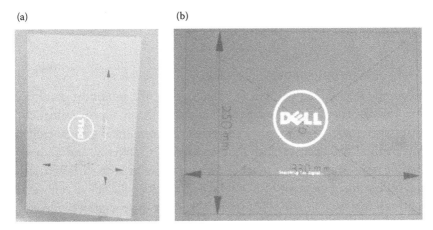

FIGURE 8.6 3D system design and testing. (a) The whole system design based on the geometric constraint. (b) Testing system before sending the design for fabrication.

Our prior study [167] found that it is difficult to achieve high calibration accuracy if the calibration features are either too small or too large: too small feature size usually results in larger calibration error due to sampling, and too large feature size leads to less accuracy calibration for lens distortions. Our experience tells that using a checker square size of 20 pixels or more is always advisable, and the whole FOV is recommended to have at least 100 feature points for optimization. In this study, we use 147 number of circles on the calibration target.

To fabricate a calibration target, one can purchase a calibration target commercially available. However, those calibration target is usually very expensive. If one is only interested in achieving good accuracy, he/she can actually make the calibration target by him/herself. The calibration target making starts with designing the calibration target using AutoCAD or some high-precision design software to generate the desired dimension and number of feature points. For the particular target we use, one can also use OpenCV to generate such a target image. Then use a high-resolution printer to print the digital version at preferably 1:1 ratio. Find a large enough size picture frame, and buy some spray glue to spray glue on the picture glass uniformly. Finally, glue the printed version onto the glass. Once it dries out, the calibration target can be used for calibration.

8.4 SYSTEM CALIBRATION

System calibration includes the nonlinear gamma calibration and the geometric properties (e.g., focal length of the lens and transformation between camera and projector) calibration. This section discusses both calibrations.

8.4.1 Nonlinear Gamma Calibration

Before the system can be used for 3D imaging, the nonlinear gamma of the projector should be properly calibrated and the nonlinear influence of the projector should be properly corrected. The nonlinear gamma calibration method has been extensively discussed in Chapter 6, and an active method is recommended in practice. Our extensive experience tells that the grayscale value increment of 5 is sufficient to adopt the active calibration approach that we discussed. Therefore, the user can generate a sequence of full-size images with grayscale values ranging from $0, 5, 10, \ldots, 255$, and send them to the projector. The projector projects these images one by one to a white board, and the camera captures these images sequentially. To alleviate the influence of noise, when the grayscale value of 255 image is projected, the camera aperture should be properly adjusted such that the captured images should be close to be saturated but not saturated. Figure 8.7a illustrates a good gamma calibration curve.

It should be noted that some projectors (actually most of the projectors we have used) may not fully respond to the grayscale image value ranging from 0 to 255. Because the gamma curve should be monotonic, if the projected image intensity does not change when the input image intensity changes, those grayscale values cannot be used. Therefore, in practice, the full range of usable grayscale values that can be used is smaller. Figure 8.7b

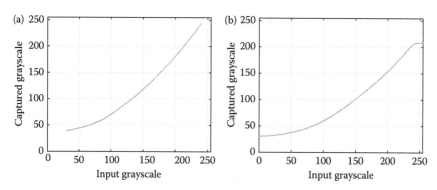

FIGURE 8.7 Good versus bad gamma curve. (a) A good curve that does not have any flat area in both ends and occupies as large range as possible. (b) Full range input grayscale value cannot be used because the projector does not respond when the input grayscale value is large than 240.

illustrates a case when the grayscale values of 240–255 cannot be used for practical measurements. For this case, the recommended grayscale value range is 0–240.

8.4.2 Geometric Calibration

Once the nonlinear gamma are calibrated, they can be used to generate ideal sinusoidal patterns for system calibration following the procedures described in Chapter 7. The temporal phase unwrapping is preferable for geometric calibration. To achieve high-accuracy mapping, it is also desirable to use more steps and narrower fringe patterns to generate the mapping between the projector and the camera images. In this research, we used a nine-step phase-shifting algorithm to generate phase-shifted fringe patterns, and the fringe pitch (number of pixels per fringe period) of 18 pixels. Instead of using multifrequency phase shifting, we used the simple coding method for temporal phase unwrapping.

After capturing a number of different poses of the calibration target, we can use OpenCV calibration toolbox to estimate system parameters. It is important to keep in mind that the calibration target orientations should be symmetric relative to the system to avoid calibration bias. In other words, if one rotates the calibration target to left, one should rotate to right, to up, and to down approximately the same angle. Figure 8.8 shows five representative poses we used for our system calibration.

Using 11 different calibration poses, we estimated the camera parameters as

$$A^c = \begin{bmatrix} 1471.59 & 0 & 315.83 \\ 0 & 1464.40 & 197.54 \\ 0 & 0 & 1 \end{bmatrix} \text{pixels}, \tag{8.11}$$

$$R^c = \begin{bmatrix} 1 & 0 & 0 \\ 0 & 1 & 0 \\ 0 & 0 & 1 \end{bmatrix}, \tag{8.12}$$

$$T^c = \begin{bmatrix} 0 \\ 0 \\ 0 \end{bmatrix} \text{mm}. \tag{8.13}$$

Here A^c is the intrinsic parameter matrix for the camera, R^c is the extrinsic parameter rotation matrix for the camera, and T^c is the extrinsic translation vector for the camera. It can be seen that we aligned our coordinate system with the camera lens and thus no rotation or translation for the camera

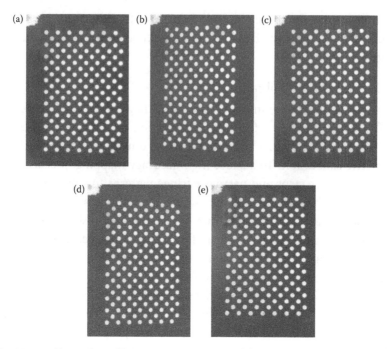

FIGURE 8.8 Example calibration orientations. (a) Rotate up. (b) Rotate left. (c) Middle. (d) Rotate right. (e) Rotate down.

lens coordinate system. Here also noted that A^c obtained from camera calibration software (i.e., OpenCV for this case) is in camera pixel, one can verify the focal length by considering the pixel size, for example,

$$f_u^c = 1471.59 \times 5.6 \, \mu m = 8241 \, \mu m = 8.241 \, mm.$$

Apparently, the focal length is very close to the focal length of the lens that is attached to the camera (i.e., 8 mm). It is important to note that the calibration software only provides the *effective* focal length (i.e., the pupil center to the camera imaging sensor plane) that is not necessarily the actual focal length of the lens itself. About 8.241 mm is close enough to the focal length of the lens, and one can believe that this calibration software provides reasonable estimates.

We also estimated projector parameters as

$$A^p = \begin{bmatrix} 1943.46 & 0 & 629.60 \\ 0 & 1933.62 & -4.68 \\ 0 & 0 & 1 \end{bmatrix} \text{pixels}, \tag{8.14}$$

$$R^p = \begin{bmatrix} 1.0000 & -0.0044 & 0.0068 \\ 0.0047 & 0.9989 & -0.04723 \\ -0.0066 & 0.04726 & 0.9987 \end{bmatrix}, \tag{8.15}$$

$$T^p = \begin{bmatrix} -1.03 \\ 170.69 \\ 0.69 \end{bmatrix} \text{mm}. \tag{8.16}$$

A^p is the intrinsic parameter matrix for the projector, R^p is the extrinsic parameter rotation matrix for the projector, and T^p is the extrinsic translation vector for the projector. Again, here the projector intrinsic matrix is in pixels, and the focal length in millimeter is

$$f_u^p = 1943.46 \times 7.6 \, \mu m = 14770 \, \mu m = 14.770 \, mm,$$

which is also close to the nominal focal length of the projector lens (14.95 mm). One may be curious about the -4.68 pixel in the A^p matrix. This negative principle point indicates that the optical axis actually intersects a point that is outside of the DMD chip, and this is a result of off-axis projection mechanism of the projector.

The translation vector tells the distance between the camera lens and the projector lens, one can measure the distance, roughly between them to verify these numbers. Once all the numbers are roughly correct, one can then use these numbers for 3D reconstruction.

Finally, one may notice that the nonlinear distortions for both camera and projector lenses were ignored for this example to illustrate the procedures of developing a practical system. More accurate 3D imaging system could be developed by considering the nonlinear distortion of the lenses.

8.5 3D RECONSTRUCTION

Once the system is calibrated, one can perform measurement to further visually verify the success of the developed system. As an example, one can reconstruct 3D geometry of the calibration target using one of the pose fringe images. Figure 8.9 shows an example of reconstructing 3D geometry for one pose. Figure 8.9b shows that the reconstructed 3D surface is roughly flat, indicating that the calibration might work reasonably well. It should be noted that the bumps on the 3D surface was introduced by the large noise in the black area of the calibration target.

More complex 3D shape can also be recovered using the calibrated system. As a example, we measured a complex statue: Zeus bust, as shown

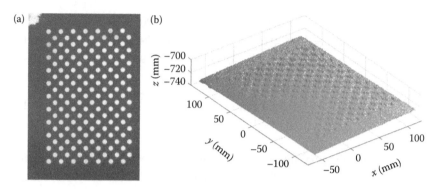

FIGURE 8.9 Example of reconstructing 3D surface of one calibration orientation. (a) Image of the calibration target. (b) 3D reconstruction.

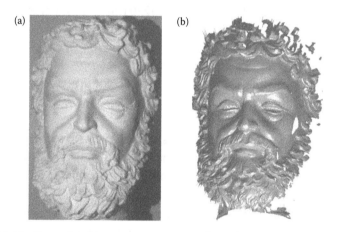

FIGURE 8.10 Example of reconstructing 3D surface of a complex statue. (a) Image of the statue. (b) 3D reconstruction.

in Figure 8.10. Once, again, the 3D reconstructed shape visually matches with the real statue fairly well, further confirming the calibration works reasonably well.

More quantitative verifications can be carried out by measuring standard object (e.g., a sphere) and comparing the measured data with the real geometry. As such a verification has been detailed in Chapter 7, we do not present such data again in this chapter.

8.6 SUMMARY REMARKS

This chapter has presented a practical 3D imaging system development. The hardware design and development require a number of iterations to achieve the desired specifications. System calibration including both non-linear gamma calibration and the geometric parameter calibration. Once the system is integrated and calibrated, one can perform some experiments to further verify the performance of the system. It is important to note that all the numerical numbers we used in this chapter were based on our experience and used as illustrations, the reader is encouraged to learn from his/her own experience to develop systems that meet the needs.

CHAPTER **9**

Practical Considerations for System Development

O PTIMAL DESIGN IS ALWAYS a difficult yet vital task for any hardware system development, often under certain constraints. 3D imaging system design is no exception, a good design can achieve high-quality measurement, and a bad design can be costly yet without achieving its potential capabilities. This chapter discusses some of the factors we think are important to 3D imaging system development. We will mainly focus on addressing one key factor (i.e., projected fringe angle) that since this factor has not been well recognized in the community.

9.1 ANGLE OF FRINGE STRIPES

To our knowledge, one of the vital factors that not many in the field actually realize is the projected fringe angle that could substantially impact measurement quality. Given a designed hardware system, the common practice is to use either horizontal or vertical or both directional fringe patterns, assuming that measurement quality will be automatically ensured. However, it is well known that the design of such a system is usually not easy and often involves complicated trial-and-error procedures. The optimal design essentially will provide the best sensitivity to depth variation. In other words, the phase changes are the largest for a given depth variation (i.e., maximize $\partial\Phi/\partial z$).

Assume the phase error caused by the system noise is $\delta\Phi_e$, which is the same once the system is set up for a given phase-shifted fringe patterns. For a simple reference-plane-based calibration method, the relationship between the depth z and the phase difference is $z = z_0 + c \times \Delta\Phi$ [67], here z_0 is the constant shift, and c is the calibration constant. Therefore, the depth error caused by the system noise is approximately $\Delta z_e = c \times \delta\Phi_e$. This indicates that the larger the calibration constant, the larger the measurement error will be induced by noise (i.e., the lower the measurement accuracy will be achieved). When the phase sensitivity ($\partial\Phi/\partial z$) is higher, the calibration constant c is smaller, and thus depth measurement accuracy is higher.

Before Wang and Zhang's work [166], no prior attention had been paid toward orienting the projected fringe patterns such that the system can achieve the optimal performance. This section reveals Wang and Zhang's finding [166] that the fringe angle plays a vital role in determining the optimal performance of the 3D imaging system with a DFP technique, and simply projecting horizontal or vertical fringe patterns may not be the best option. It becomes crucial to determine the optimal fringe angle for a given system setup without changing its mechanical design. This section presents an efficient method to determine the optimal fringe angle. Specifically, by projecting a set of horizontal and vertical fringe patterns onto a step-height object, we obtain two phase maps, from which the phase differences between the top and the bottom surface of the step-height objects can be calculated. Finally, the mathematical vector operation on these phase differences can be used to determine the optimal projection angle. Our further study indicated that if the projected fringe stripes have the optimal angle, the phase is the most sensitive to the depth variations for a given system setup. In contrast, if the fringe stripes are perpendicular to the optimal fringe direction, the system is the least sensitive to the depth variations. Here the optimal performance, again, means that the system is the most sensitive to the depth changes of the object surface, i.e., $\partial\Phi/\partial z$ is the largest for a given setup. This section also presents an example showing that using both horizontal and vertical fringe patterns together for 3D reconstruction may not bring any benefits, if not jeopardize the quality.

To achieve high sensitivity, one can adjust the relative position and orientation between the projector and the camera, which is usually not easy. Instead of mechanically redesigning the system, Wang and Zhang [166] proposed to change the projected fringe stripe orientation such that the system can achieve the best sensitivity. Specifically, the following procedures can be used to determine the optimal fringe angle:

Step 1: *Pattern projection.* Project horizontal and vertical fringe patterns onto a step-height object and a uniform flat reference plane; and obtain four absolute phase maps by employing a multifrequency phase-shifting algorithm: horizontal absolute phase map of the object Φ_{Ho}, vertical absolute phase map of the object Φ_{Vo}, horizontal absolute phase map of the reference plane Φ_{Hr}, and vertical absolute phase map of the reference plane Φ_{Vr}.

Step 2: *Phase retrieval.* Calculate the difference phase maps by subtracting the object phase maps with the corresponding reference phase maps:

$$\Delta\Phi_{Hd} = \Phi_{Ho} - \Phi_{Hr}, \tag{9.1}$$

$$\Delta\Phi_{Vd} = \Phi_{Vo} - \Phi_{Vr}. \tag{9.2}$$

Step 3: *Phase difference calculation.* Compute the phase difference between top and bottom surfaces of the step object using the following equations:

$$\Delta\Phi_{H} = \Delta\Phi_{Hd}^{t} - \Delta\Phi_{Hd}^{b}, \tag{9.3}$$

$$\Delta\Phi_{V} = \Delta\Phi_{Vd}^{t} - \Delta\Phi_{Vd}^{b}. \tag{9.4}$$

Step 4: *Optimal fringe angle determination.* Determine the optimal fringe angle using the following equation, which is illustrated in Figure 9.1.

$$\theta_{o} = \tan^{-1}\left[\Delta\Phi_{V}/\Delta\Phi_{H}\right]. \tag{9.5}$$

This is essentially the angle of the vector $\vec{v} = \Delta\Phi_{H}\vec{i} + \Delta\Phi_{V}\vec{j}$. Here $\vec{i}$ and $\vec{j}$ are the unit vector along x and y axes, respectively.

As has been explained previously, if the projected fringe patterns use the optimal angle θ_{o}, the phase changes the greatest with the same amount of depth variation. Therefore, such a system can be used to measure the smallest features on an object surface for a given hardware system configuration. This finding is especially valuable for a DFP system where the orienting of the projected fringe patterns can be easily realized.

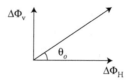

FIGURE 9.1 Calculation of the optimal fringe angle. (From Y. Wang and S. Zhang, *Appl. Opt.* 52(29), 7094–7098, 2013. With permission.)

It should be noted that the projected patterns generated by a computer can be perfectly horizontal or vertical, but the captured fringe patterns by the camera may have small angle error (i.e., the fringe patterns may not be perfectly horizontal or vertical). The optimal fringe angle determination method was based on the computer-generated fringe patterns before sending to the projector, and thus the angle error can be negligible. Practically, if the projected fringe patterns are close to optimal angle, the phase change is nearly the largest with a given depth variation. In contrast, if the fringe angle is close to being perpendicular to optimal angle, the phase change is close to being zero with the same depth change, meaning that the measurement sensitivity is very low and the measurement accuracy is drastically jeopardized due to factors such as system noise and/or phase error.

The optimal angle determination method was tested on a DFP system. The system includes a DLP projector (Samsung SP-P310MEMX) and a digital CCD camera (Jai Pulnix TM-6740CL, San Jose, California). The camera uses a 16-mm focal length megapixel lens (Computar M1614-MP, Cary, North Carolina). The camera resolution is 640 × 480, with a maximum frame rate of 200 frames/s. The projector has a resolution of 800 × 600 with a projection distance of 0.49–2.80 m.

A standard step-height block was used to determine the optimal angle of our DFP system. The block size is approximately 40 mm (H) × 40 mm (W) × 40 mm (D). Figure 9.2 shows the measurement results using horizontal and vertical fringe patterns. Figure 9.2a and d, respectively, shows one of the horizontal and vertical fringe images captured by the camera. Figure 9.2b and e shows the corresponding phase difference maps (Φ_{Hd} and Φ_{Vd}), which were obtained using Equations 9.1 and 9.2. To better illustrate the phase difference maps, Figure 9.2c and f, respectively, shows the same cross section of the phase map shown in Figure 9.2b and e.

Taking the difference between the top surface of the block and the bottom surface of the block, we can obtain $\Delta\Phi_H$ and $\Delta\Phi_V$, from which we can determine the optimal angle (θ_o) using Equation 9.5. It should be

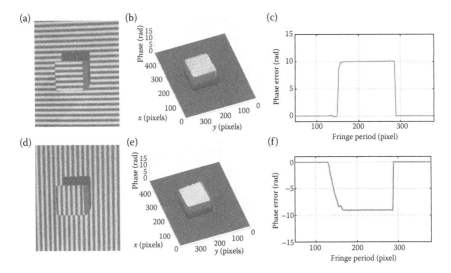

FIGURE 9.2 Phase measurements using horizontal and vertical fringe patterns. (a) One of the captured horizontal fringe patterns. (b) Phase difference map Φ_{Hd}. (c) 250th row cross section of (b). (d) One of the captured horizontal fringe patterns. (e) Phase difference map Φ_{Vd}. (f) 250th row cross section of (e). (From Y. Wang and S. Zhang, *Appl. Opt.* 52(29), 7094–7098, 2013. With permission.)

noted that the averaged phase values should be used for the phase differ-ence determination on the step-height object to alleviate the noise effect. Namely, the phase maps of a small area on the top surface and that on the bottom surface were averaged to calculate the phase difference for each fringe angle. In this case, the optimal fringe angle is approximately $\theta_o = -0.73$ rad. In contrast, if the fringe stripe is perpendicular to the optimal fringe stripe direction, the system is the least sensitive to the depth changes. In other words, if the fringe angle is ($\theta = 0.84$ rad), the phase difference map of the step-height block should be close to zero.

The optimal angle determination idea was then verified through exper-iments. Figure 9.3 shows the results. Figure 9.3a and d, respectively, shows one of the captured fringe images under the worst and optimal fringe angles. Figure 9.3b and e shows the corresponding phase difference maps. The cross sections are shown in Figure 9.3c and f. These experiments show that the phase difference is indeed close to zero if the fringe direction is per-pendicular to optimal fringe direction; and when the projected patterns use the optimal fringe angle, the phase difference is drastically larger than either the horizontal or the vertical fringe patterns that are commonly used, as illustrated in Figure 9.2. It should be noted that during all experiments,

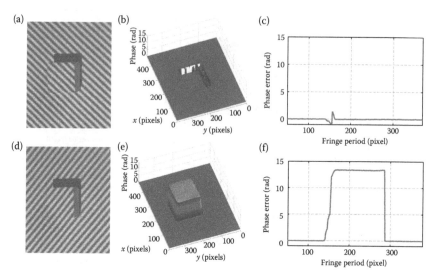

FIGURE 9.3   Results for the fringe patterns with the worst and the optimal fringe angles. (a) One of the captured fringe images with θ = 0.84 rad, the worst fringe angle. (b) Phase difference map of (a). (c) 250th row cross section of (b). (d) One of the captured fringe images with $θ_o$ = −0.73 rad, the optimal fringe angle. (e) Phase difference map of (d). (f) 250th row cross section of (e). (From Y. Wang and S. Zhang, *Appl. Opt.* 52(29), 7094–7098, 2013. With permission.)

the whole hardware system remained untouched, the object was positioned the same location, and the fringe period was the same. These data demonstrated that the optimal fringe angle can be determined, and the system is the most sensitive to the depth variation under the optimal condition.

A more complex sculpture was also measured using the optimal fringe angle and the worst fringe angle, as shown in Figure 9.4. Figure 9.4a and d shows one of the captured fringe patterns and Figure 9.4b and e shows the corresponding phase difference maps. It can be seen that the phase difference map is nearly flat and no details are obvious on Figure 9.4b when the worst fringe angle is used. In contrast, Figure 9.4e clearly shows the highly detailed features on the difference map when the fringe patterns use the optimal fringe angle. The phase difference maps were further converted to depth maps using the simple phase-to-height conversion algorithm discussed in Reference 67. The depth scaling coefficient was obtained using optimal fringe angle, and applied to both phase difference maps. Figure 9.4c and f, respectively, shows the worst and the best result. It clearly shows that using the optimal fringe angle, the 3D object can be properly measured with details. Yet, if the fringe orientation rotates 90°, 3D shape cannot be

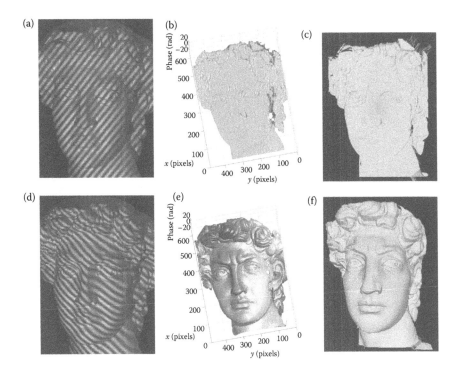

FIGURE 9.4 Sculpture results under different fringe angles. (a) One of the captured fringe patterns with the worst fringe angle $\theta = 0.84$ rad. (b) Phase difference map ($\theta = 0.84$ rad). (c) Recovered 3D shape ($\theta = 0.84$ rad). (d) One of the captured fringe patterns with the optimal fringe angle $\theta_o = -0.73$ rad. (e) Phase difference map ($\theta_o = -0.73$ rad). (f) Recovered 3D shape ($\theta_o = -0.73$ rad). (From Y. Wang and S. Zhang, *Appl. Opt.* 52(29), 7094–7098, 2013. With permission.)

properly recovered at all since the phase difference map is close to be zero across the whole range.

For a well-designed system, such as the one illustrated in Chapter 8, only horizontal or vertical fringe patterns is sufficient for high-quality 3D imaging; and the fringe orientation must be properly used, otherwise, no reasonable good quality 3D results will be obtained. For example, if we use horizontal fringe patterns, instead of vertical patterns, to measure the same object shown in Figure 8.10, we only generate the results shown in Figure 9.5a. Carefully examine the captured horizontal fringe patterns, as illustrated in Figure 9.5b, one may notice that the fringe stripes were not distorted due to surface geometry, and thus no triangulation can be used for 3D shape reconstruction. In contrast, if vertical fringe patterns are used, the fringe stripes are distorted, shown in Figure 9.5c, and thus good 3D shape can be properly reconstructed.

(a)　　　　　　　　　　(b)　　　　　　　　　　(c)

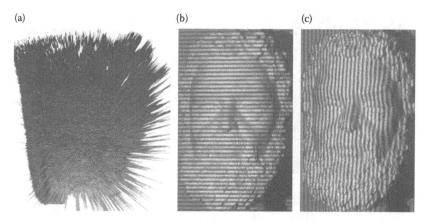

FIGURE 9.5 Measuring a 3D statue with incorrect fringe orientation for a well-designed system cannot generate good results. (a) 3D reconstructed shape using horizontal fringe patterns. (b) One of the phase-shifted horizontal patterns. (c) One of the phase-shifted vertical patterns.

Furthermore, using both horizontal and vertical fringe patterns together for 3D reconstruction intuitively should provide better results since more information are used. However, for a well-designed system, this common practice may not bring any benefits. Figure 9.6 shows the comparing results. There are no obvious differences between the result solely using vertical fringe patterns and that using both horizontal and vertical fringe patterns. In fact, the result from the latter approach actually have larger noise than the result from the former. Therefore, practically, if a system is well designed, only one single orientation fringe pattern is necessary for 3D shape reconstruction.

(a)　　　　　　(b)　　　　　　(c)　　　　　　(d)

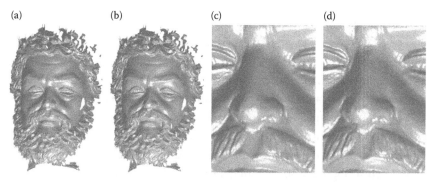

FIGURE 9.6 For a well-designed system, using both orientation fringe patterns for 3D reconstruction may not yield better results. (a) 3D shape only using vertical fringe patterns. (b) 3D result using both horizontal and vertical fringe patterns. (c) Zoom-in view of (a). (d) Zoom-in view of (b).

Finally, one can prove that the optimal angle is fundamentally based on epipolar geometry: if the fringe angle is perfectly perpendicular to the epipolar line direction pixel by pixel, then the highest possible sensitivity will be achieved. Because projector nonlinearity is usually not significant, the local optimal angle is almost identical and thus a single angle, as demonstrated in this work, can be used globally for the whole projected pattern.

9.2 OTHER CONSIDERATIONS

There are other factors that should be considered to optimize 3D imaging system design, this section presents some major factors, angle between projector and camera, fringe density, number of patterns, and hardware components.

9.2.1 Angle between Projector and Camera

Apparently, the angle between projector and the camera plays a crucial role because this is a triangulation-based 3D imaging system. In general, the larger the angle ($<90°$), the higher the achievable sensitivity yet the larger the shadow/occlusion problem caused by projector/camera. Therefore, one should chose a proper angle between projector and camera to compromise sensitivity with sensing area. Our experience tells usually a $12°–15°$ is fairly good. Figure 9.7 presents the results if the angle between projector and camera is too large or too small. Apparently, neither measurement is desirable. One may notice that if the angle is too small, the fringe stripe

(a)   (b)   (c)   (d)

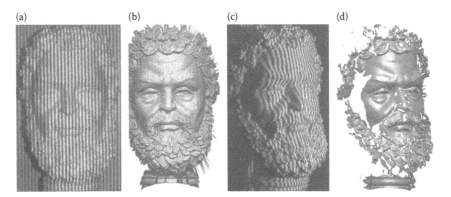

FIGURE 9.7 **(See color insert.)** 3D results when the angle between projector and camera is either too large or too small. (a) One of the fringe patterns when the angle is too small. (b) 3D result when the angle is too small. (c) Fringe pitch = 72 pixels. (d) 3D result when the angle is too large.

distortions caused by surface geometry is very small (Figure 9.7a) and thus 3D result is very noisy (Figure 9.7b). In contrast, if the angle is too large, the fringe stripes are significantly distorted by surface geometry (Figure 9.7c). For those area that measurement can be performed, the measurement quality is extremely high. However, the problem is very obvious: there are a lot of areas that cannot be measured (Figure 9.7d).

9.2.2 Fringe Density

It is well known that the denser the used patterns, the higher the achievable sensitivity. However, because both camera and projector are discrete devices, the number of pixels to represent one fringe must be a certain number (e.g., >10 pixels). Otherwise, the projected and/or captured fringe contrast is not be high (or SNR is low), and thus the measurement quality is low. In contrast, if fringe stripes are too wide, the random noise is significant. Figure 9.8 presents the results if fringe stripes are too dense or too

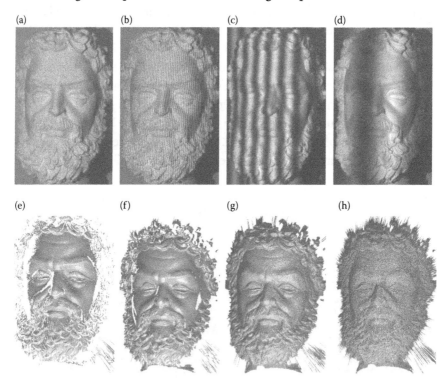

FIGURE 9.8 **(See color insert.)** 3D reconstructions of using different fringe density. (a) Fringe pitch = 6 pixels. (b) Fringe pitch = 9 pixels. (c) Fringe pitch = 72 pixels. (d) Fringe pitch = 300 pixels. (e)–(h) 3D reconstruction using the three above phase-shifted fringe patterns and temporal phase unwrapping.

sparse, again, neither yields good quality measurement. Figure 9.8e and f shows 3D results from too dense fringe patterns. One can clearly see that there are a lot area that cannot be measured due to very low fringe contrast, and measured surface has some obvious high-frequency stripes that are a result of low-quality fringe pattern representations. Figure 9.8g and h shows results from too wide fringe stripes. Clearly, the random noise is larger when fringe stripes are wider.

9.2.3 Pattern Number

Due to the averaging effect, the more patterns used, the higher measurement accuracy could be achieved. However, the slower the measurement will be. Therefore, there is always a trade-off between speed and accuracy in practice. Figure 9.9 illustrates the progression of using an increased number of fringe patterns, as can been seen here. It shows that using too few patterns (e.g., three) will have larger noise, and using too many patterns (e.g., 18 vs. 9) may not be necessary since the improvement is substantial.

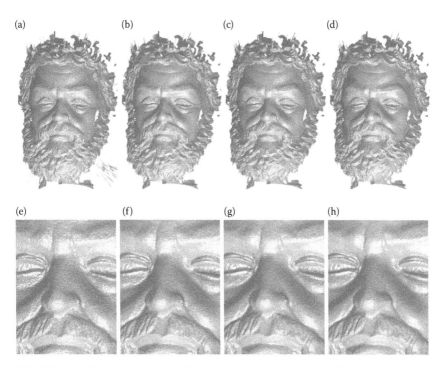

FIGURE 9.9 **(See color insert.)** 3D reconstructions of using different number of phase-shifted fringe patterns. (a) Three step. (b) Six step. (c) Nine step. (d) 18 step. (e)–(h) Zoom-in view of above results.

9.2.4 Capture Exposure

Although phase-shifting algorithm is not very sensitive to captured fringe pattern brightness, it is still very important to capture as good quality fringe patterns as possible to ensure high-quality 3D reconstructions especially a small number of fringe patterns are used and the fringe stripes are not too narrow. A good exposure typically refers to the captured fringe patterns that are as bright as possible but not saturated. Figure 9.10 shows some example frames that illustrates the influence of fringe patterns when the number of step is three and fringe pitch is 36 pixels. As can be seen from this example, if the fringe patterns are saturated (Figure 9.10a), one can stripe the error along fringe direction on recovered 3D geometry (Figure 9.10e); and if the exposure is too low (Figure 9.10d), large random noise presents 3D geometry (Figure 9.10h).

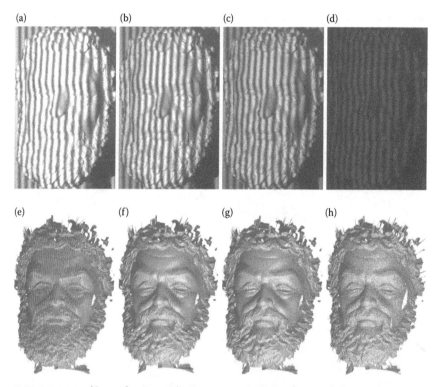

FIGURE 9.10 **(See color insert.)** 3D reconstructions of using different exposures. (a) Too much exposure fringe pattern. (b) Slightly too much exposure fringe pattern. (c) Good exposure fringe pattern. (d) Too low exposure. (e)–(h) 3D reconstruction using the three above phase-shifted fringe patterns and temporal phase unwrapping.

9.2.5 Hardware Components

Finally, it is obvious that the better the hardware (i.e., camera and projector) used for a 3D imaging system, the higher the measurement quality should be generated. However, the more expensive the system is. Therefore, developing a cost-effective 3D imaging system is always challenging since it requires one to have knowledge of commercially available hardware components on the market.

9.3 SUMMARY REMARKS

This chapter discussed some of the key factors for 3D imaging system optimization. The chapter spent most of its space on the optimal fringe angle determination for a given system, and briefly discussed some well-known factors. The reason of focusing on the optimal angle is that this is a relatively ignored factor that could be substantially impact measurement quality, if not properly used. Optimization is always a big topic for any system design, truly optimized system could involve a lot more than we discussed in this chapter. Therefore, sometimes one has to learn from his/her experiences to design the best possible system given all constraints imposed by applications.

CHAPTER 10

Toward High-Speed 3D Imaging with Phase-Shifting Methods

O NCE THE 3D IMAGING SYSTEM is developed, the next stage is to increase its speeds. The measurement speed improvement involves both software and hardware, yet is fundamentally limited by hardware. Therefore, this chapter focuses on how to fundamentally achieve the highest-possible 3D imaging speed provided that the software processing can catch up with hardware. Specifically, this chapter summarizes a decade long research efforts toward achieving high-speed 3D imaging leveraging the DMD platforms. We cover two milestones that have been reached over the course of more than a decade: the first milestone was to achieve high-resolution real-time 3D imaging technology at video frame rate (e.g., 30 Hz or better) including simultaneous 3D shape acquisition, reconstruction and display; and the second milestone was to develop the digital binary defocusing technology to perform 3D imaging at a rate of tens of kHz.

10.1 INTRODUCTION

Recent advancements in 3D computational methods have led to rapid growth in the field high-resolution, real-time 3D imaging over the past decades, and numerous techniques have been developed to recover 3D shape based on different principles [19,168]. The most popular methods include the TOF, laser triangulation, shape from focus and defocus, stereo vision, structured light, and DFP, which have been discussed in Chapter 1, and summarized in the book by Zhang [1].

The structured-light method became one of the most important 3D imaging technologies for both scientific research and industrial practices mainly because of its simplicity and speed [17]. Real-time to high-speed 3D imaging has become even more popular because the processing power of a regular, modern computer (even a tablet) can handle such large amounts of data [21]. In our view, *real-time 3D imaging* includes three major components: acquisition, reconstruction, and redisplay all occur at speeds of 24 Hz or higher.

Even though there are real-time 3D imaging techniques developed based on other principles, such as TOF [169], active stereo vision [20], and structured light [170], and the DFP techniques stand out since it enables one to achieve camera pixel spatial resolution.

This chapter mainly focuses on the research efforts toward high-speed 3D imaging using the DMD platforms. First, we present the first-ever high-resolution real-time 3D imaging system using the DFP method that was developed: the system can simultaneously acquire, reconstruct, and redisplay 3D shapes at 30 Hz with over 300,000 measurement points per frame [62]. We then present some recent innovations on the digital binary phase-shifting technique [64] to achieve kHz rate 3D imaging rate [171]. Realizing the fundamental limitations of the binary defocusing technique (e.g., the squared binary method has smaller measurement depth range than the conventional method), researchers have developed methods to improve the measurement accuracy without sacrificing the measurement speed; and also developed methods to increase the measurement depth range without losing measurement quality. This chapter summarizes the developments of the optimal pulse width modulation (OPWM) [172], the binary dithering/half toning technique [173], and the optimizations of dithering technique [174,175].

This chapter serves as a sort of overview of research efforts toward high-speed 3D imaging. From our own experience and perspective, we address how the real-time 3D imaging technologies evolve by leveraging the unique features of hardware technologies. We hope that this chapter provides readers with a coherent piece of literature on the status of the most recent high-speed 3D imaging technologies, motivates them to further advance these technologies, and drives them to adapt these technologies for their specific applications.

Section 10.2 briefly presents the first real-time 3D imaging system that was developed by modifying the commercially available DLP projectors. Section 10.3 presents the recent advancements on superfast 3D imaging technologies. Section 10.4 discusses some challenging problems to conquer to increase the value of real-time 3D imaging technologies in practical applications. Finally, Section 10.5 summarizes this chapter as well as the whole book.

10.2 REAL-TIME 3D IMAGING WITH SINUSOIDAL FRINGE PROJECTION

Great efforts toward superfast 3D imaging started with modifying a DLP projector mainly because of the availability of hardware at the time and the limitations of the other hardware components. The commercially available DLP projectors typically generate color images that are fed to the projectors. The colors are generated by putting a rapidly spinning color wheel into the optical path and properly synchronizing the color wheel with the computer-generated signal. The synchronization signal is usually picked up by a photosensor behind the color wheel with some associated electronics.

As discussed in Chapter 3, DLP works off the time modulation to generate grayscale images, and thus synchronizing the projector and the camera becomes extremely important to properly capture the desired fringe images used for high-quality 3D imaging. Moreover, using color fringe patterns is not desirable for high-quality 3D imaging [92], so the elimination of color is vital. Furthermore, the commercially available DLP projectors are typically nonlinear devices to accommodate for human vision, but generating ideal sinusoidal patterns is required for high-quality 3D imaging. Meanwhile, the unique projection mechanism of DLP technology, sequentially projecting red, green, and blue channels, permits the natural way of high-speed 3D imaging if a three-step phase-shifting algorithm is used. In summary, to achieve high-speed 3D imaging at a desired high quality with a commercially available DLP projector, the following hardware challenges need to be conquered: (1) elimination of the color wheel from the optical path; (2) precise synchronization between the camera and the projector; and (3) corrections for the nonlinear gamma effects.

To conquer the aforementioned three challenges, modifying the commercially available digital video projector is required. The most important modifications made were removing the color filters of the color wheel and supplying an external trigger signal such that the projector can still take the regular color image from the computer and project the red, green, and blue channels sequentially, albeit in monochromic mode. The external trigger signal was generated with microcontroller to mimic the signal from the photodetector such that the projector cannot tell whether the signal was from the photodetector or the circuit and thus behaves independently of the signal source.

After these modifications, the projector projects a monochrome fringe image for each of the RGB channels sequentially. Each "frame" of the projected image is actually three separate images. By removing the color wheel and placing each fringe image in a separate channel, the projector can produce three fringe images at 80 Hz (240 individual color channel refresh rate).

It is important to note that, as indicated in Figure 10.1a, the timing of different color channels is different (red channel lasts the longest time and blue channel lasts the shortest time) for the modified digital video projector, making the precise synchronization between the projector and the camera more difficult because rapidly changing the camera exposure time from one frame to the next was not permitted. The synchronization between the projector and the camera requires a time-consuming trial-and-error process. It is also important to note that the projector also has a clear channel that is intended to increase the overall brightness of the projector, but this does not provide additional useful information for 3D imaging purpose.

As addressed in Chapter 2, three fringe images can be used to reconstruct one 3D shape if a three-step phase-shifting algorithm is used. This perfectly fits into the DLP technology schema where three patterns can be encoded into each of the three primary color channels of the projector. Since color fringe patterns are not desirable for 3D imaging, Zhang and Huang [62] developed a real-time 3D imaging system based on a single-chip DLP projector and white light technique. Figure 10.2 shows the system layout. The computer-generated color-encoded fringe image is sent to a single-chip DLP projector that projects three color channels sequentially and repeatedly in grayscale onto the object. The camera, precisely synchronized with projector, is used to capture three individual channels separately and quickly. By applying the three-step phase-shifting algorithm to three fringe images, the 3D geometry can be recovered. Averaging three fringe images will result in a texture image that can be further mapped onto the recovered 3D shape to enhance certain visual effects. Since three fringe images are sufficient to recover one 3D shape, the 3D measurement speed

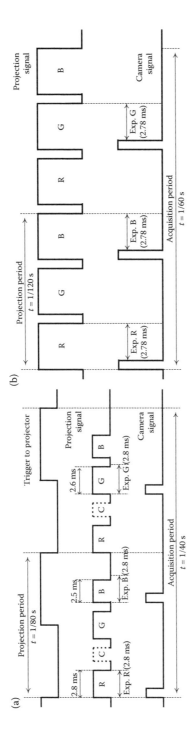

FIGURE 10.1 Timing chart of the measurement system. (a) The first projector (Kodak DP900) was modified without firmware changes. (b) The second projector (PLUS U5-632h) was modified with an additional PLUS Vision firmware update. (From T. Bell and S. Zhang, *Opt. Eng.* 53(11), 112–206, 2014. With permission.)

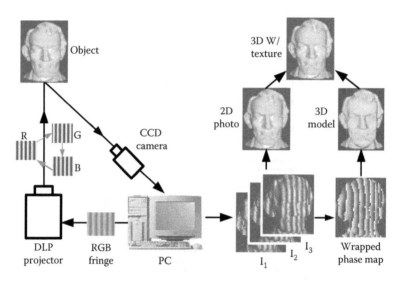

FIGURE 10.2 **(See color insert.)** Real-time 3D imaging system layout. (From S. Zhang, *Opt. Laser Eng.* 48(2), 149–158, 2010. With permission.)

is up to 80 Hz when the projector is refreshing images at 240 Hz. However, due to the speed limit of the camera used and the asynchronous data transfer time requirement, it takes two projection cycles to capture three fringe images; thus, the measurement speed is 40 Hz [62]. Figure 10.1a shows the timing chart for the real-time 3D imaging system.

As aforementioned, real-time 3D imaging involves real-time 3D shape acquisition, reconstruction, and redisplay. The processing time is tremendous since computing the wrapped phase point by point using a standard arctangent function is slow, and the spatial phase unwrapping is usually extremely time consuming. These are just some of the reasons why real-time 3D shape reconstruction was difficult back round 2004 when personal computer power was limited and GPU technology was not mature. By developing a new method for computing wrapped phase using intensity ratio instead of standard arctangent functions [176], and substantially optimizing phase-unwrapping algorithm (detailed in Chapter 5), Zhang and Huang successfully achieved real-time 3D data processing for an image resolution of 532 × 500 [177]. Figure 10.4 (later in the chapter) shows an example frame which is measuring human facial expressions in real time (40 Hz in this case).

However, there are two major issues associated with the first real-time 3D imaging system:

1. *Inaccurate calibration method.* To achieve real-time 3D data processing, the simple reference-plane-based calibration method was adopted. As a result, the recovered 3D geometry could be significantly distorted if the object is away from the reference plane, or if the hardware system is not properly configured [47].

2. *Low phase quality.* The phase quality is reasonable but not the best in principle. This is because the three color channels have different illumination times, shown in Figure 10.1a, but the camera uses the exposure time to capture these channels. The result is not all channels are properly captured.

Further hardware modifications are required in order to solve for these two problems. Fortunately, PLUS Vision agreed to do firmware modifications for one of their projector models, PLUS U5-632h. With the company's assistance, besides removing color filters, the modified projector only projects red, green, and blue channels without the clear channel, and all these channels have the same duration time. These modifications were highly desirable for high-quality 3D imaging. Furthermore, because of these modifications, the projection speed can be at 120 Hz (or 360 Hz individual channel refreshing rate), speeding up the whole measurement (to 120 Hz). Zhang and Yau [119] developed the timing chart for the second generation real-time 3D imaging system (Figure 10.1b).

Besides improved phase quality, the new system also adopted the more complex, yet more accurate calibration method that Zhang and Huang developed [47] whose principles are discussed in Chapter 7. The new calibration method requires substantial computation power to achieve real-time 3D data processing, making it almost impossible to achieve on a CPU on a regular personal computer. Fortunately, GPU technologies emerged and quickly expanded.

By taking advantage of the processing power of the GPU, 3D coordinate calculations can be performed in real time with an ordinary personal computer with an NVidia graphics card [164]. Moreover, because 3D shape data are already on the graphics card, it can be rendered immediately without any delay. Therefore, by these means, real-time 3D geometry visualization can also be realized in real time simultaneously. Also, because only the phase data, instead of the 3D coordinates and the surface normals, are transmitted to graphics card for visualization, this technique reduces the data transmission load on the graphics card significantly ($\sim$6

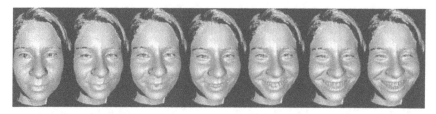

FIGURE 10.3 Examples of using the real-time 3D imaging system to capture human facial expressions. The results were captured at 60 Hz using a camera with a resolution of 640 × 480.

times smaller). In short, by utilizing the processing power of GPU for 3D coordinate calculations, real-time 3D geometry reconstruction and visualization can be performed rapidly and in real time.

Figure 10.3 shows typical results of capturing human facial expressions. The fringe images were captured at 180 Hz with a 640 × 480 image resolution, and a three-step phase-shifting algorithm was used for 3D reconstruction, meaning 3D data were captured at 60 Hz. Clearly, the facial details (even hairs) are well captured by the system, albeit the camera resolution is not very high.

Figure 10.4 shows an experimental result of measuring a live human face and rendering the results in real time. The right figure shows the real subject, and the left shows the 3D geometry acquired and rendered on the computer screen at the same time. The simultaneous 3D data acquisition,

FIGURE 10.4 Simultaneous 3D data acquisition, reconstruction, and display at 30 Hz on a regular personal computer. (From S. Zhang, *Opt. Laser Eng.* 48(2), 149–158, 2010. With permission.)

reconstruction, and display speed achieved was 30 Hz with more than 300,000 points per frame.

These technologies, though successful, require substantial projector modifications, and sometimes these modifications are impossible without the projector manufacturer's involvement. Fortunately, with about a decade of effort on high-speed 3D imaging, projector manufacturers realized the opportunities in this field of 3D imaging by producing affordable specialized projectors: LogicPD LightCommander being the first, and then the LightCrafter series by WinTech Digital. Note that the DLP Discovery platforms were available much earlier than LightCommander or LightCrafter, but they were too expensive for wide adoption.

10.3 SUPERFAST 3D IMAGING WITH BINARY DEFOCUSING

As 3D imaging technologies become more accurate and become faster, such as with the real-time systems which have been developed, the number of potential applications in the field continues to grow. These technologies have already been applied successfully to the medical field, the entertainment field, the manufacturing field, and many more. Up until now, the definition for a real-time system might be one that can capture at a rate of 30 or higher Hz. This speed can capture scenes or objects that are not too rapidly changing or moving around. A good example of this might be one's facial expression. If an object being captured does move at a very fast speed, such as a speaking mouth or a live and beating rabbit heart, 30 Hz capturing rate is not sufficient. Fortunately, the DLP platforms (e.g., DLP Discovery, DLP LightCommander, and the DLP LightCrafter) can naturally display binary images at much faster rate (e.g., kHz) though they can only display 8-bit grayscale images at a few hundred Hz. Therefore, if only binary images are necessary, 3D imaging rate can go up to kHz. As such, the binary defocusing technique [64] was invented.

The binary defocusing technique works off the simple principle of optics and instead of making the projected images focused, the projector is purposely defocused such that the images are blurry. And if the projector is defocused to a certain amount, seemingly high-quality sinusoidal fringe patterns can be generated if the squared binary images are projected. Figure 10.5 shows an example of the captured image when the projector is defocused to different amounts. It shows that when the projector is nearly focused, clear binary structures presents on the image, as shown in Figure 10.5a. With the amount of defocus increases, the binary structured

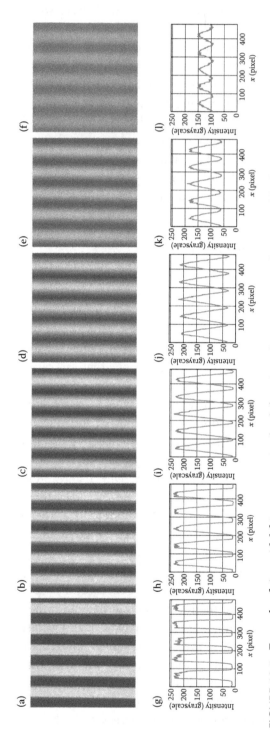

FIGURE 10.5 Example of sinusoidal fringe generation by defocusing a binary structured patterns. (a) The result when the projector is in focus. (b)–(f) The result when the projector is increasingly defocused. (g)–(l) The 240 row cross section of the corresponding above image. (From Y. Gong and S. Zhang, *Opt. Express* 18(19), 19,743–19,754, 2010. With permission.)

become less and less obvious, and when the projector is defocused to a certain amount, as shown in Figure 10.5d, seemingly good sinusoidal structures are generated. Obviously, if the projector is defocused too much, the binary structures were well blended together and no clear structures present, as shown in Figure 10.5f.

Zhang et al. [171] successfully achieved kHz range 3D imaging by using the binary defocusing technique combined with the DLP Discovery platform. Albeit fast, this new method comes with its own challenges:

- *Challenge 1: Lack of accurate calibration method.* The majority existing structured-light system calibration method assumes that the projector is at least nearly focused; and the extensively adopted easy-to-use method developed by Zhang and Huang [47] completely fails theoretically because the assumption of one-to-one mapping between the projection pixel and camera pixel cannot be established in the intensity domain.

- *Challenge 2: Low phase quality.* Due to the presence of high-frequency harmonics after defocusing, it is difficult to achieve high-quality phase especially when a small number (e.g., three or four) of phase-shifted fringe patterns are used.

- *Challenge 3: Small measurement range.* The binary patterns become high-quality sinusoidal patterns when the projector is properly defocused, and the range of this defocusing amount is relatively small [67].

- *Challenge 4: Difficulty of adopting multifrequency phase-shifting algorithms.* High-quality fringe patterns at different spatial frequencies are hard to generate simultaneously because they require different amount of defocusing [178].

To overcome challenge 1, Merner et al. [162] developed a calibration method that does not require the establishment of one-to-one mapping in intensity domain, but rather directly relating phase to depth. Such a method works well if only depth z is the primary consideration since x and y calibration accuracy is rather low. Li et al. [137] theoretically proved the existence of one-to-one mapping in phase domain regardless of the amount of defocusing, allowing adopting the existing calibration approach developed by Zhang and Huang [47] for defocused systems, which was detailed in Chapter 7.

The remaining three challenges all relate to the requirement of using properly defocused projection system, and if high-quality 3D imaging can be achieved without such a strong requirement, they can be solved all together. Therefore, to alleviate this problem, squared binary pattern is not the best option and changing the binary structures is necessary. A number of methods have been developed to address the limitations of the binary defocusing techniques. Ekstrand and Zhang [51] found that increasing the number of phase-shifted fringe patterns could reduce the error caused by high-frequency harmonics if a least-squares fringe analysis method is adopted. Moreover, since a squared wave only has odd-number harmonics without even ones, using an odd-number phase-shifting algorithm is preferable for a DFP system where the phase shift errors can be completely eliminated. Furthermore, compared with sinusoidal fringe patterns, the binary pattern has better fringe contrast, and thus has the potential to better measure microstructure [179]. However, the enhanced measurement quality was achieved by utilizing more fringe patterns, which usually means sacrificing measurement speed. For high-speed applications, this approach is not desirable.

To enhance the binary defocusing method without increasing the number of required structured patterns, modulating the squared binary patterns is probably the only option since for a conventional DFP technique, the sinusoidal fringe patterns only vary in one direction. Improving squared binary patterns becomes optimizing the square wave such that it is easier to generate sinusoidal fringe patterns through low-pass filtering. The first attempts were tried by different research groups by borrowing the pulse width modulation (PWM) method invented in the power electronics field. Fujita et al. [180] and Yoshizawa and Fijita [181] implemented the PWM technique through hardware, Ayubi et al. [182] realized the sinusoidal pulse width modulation (SPWM) technique, Wang and Zhang [172] realized the OPWM technique, and Zuo et al. [183] further improved the OPWM technique. Figure 10.6 shows some example patterns using different PWM methods. Comparing with squared binary patterns, these modified patterns looks more sinusoidal even before defocusing.

Though PWM techniques indeed improved phase quality, they all still face one problem: cannot generate high-quality pattern when the fringe stripes are too wide [66] since they fundamentally modify the patterns only in one dimension, either x or y. Since fringe patterns are 2D in nature, modifying the pattern in two dimensions (both x and y) could lead to better results. This drove the development of local area modulation methods

(a)   (b)

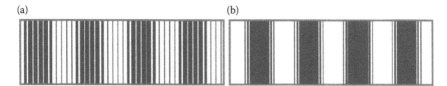

FIGURE 10.6 Example of binary patterns with different pulse width modulation method. (a) Sinusoidal pulse with modulation (SPWM) method. (b) Optimal pulse width modulation (OPWM) method. (From B. Li et al., *Opt. Laser Eng.* 54, 236–246, 2014. With permission.)

[184,185]. They certainly improved fringe quality, yet their improvements are still limited. It turns out that a great amount of research has been carried out in the printing field to represent high bits images with binary images, and such a method called dithering or half toning. The dithering techniques have been around since the 1960s [186]. The most popular methods are the random dithering [187], the simple Bayer-ordered dithering technique [188], and the error-diffusion technique [189–191].

Figure 10.7 shows the dithering results of 8-bit grayscale images with different dithering method. Apparently, the error-diffusion dithering method works better than Bayer dithering. However, Bayer dithering is still quite useful because its potential parallel processing nature.

Wang and Zhang borrowed both Bayer dithering [173] and error-diffusion techniques [68] to the field of 3D imaging. The dithering techniques substantially expand the capability of binary defocusing technique, since they can generate a large range of fringe breadths when the projector is not too much defocused, and thus the depth range of measurement is also extended. Combining the OPWM technique and dithering method, Wang et al. have developed a superfast 3D imaging system that can capture live rabbit hearts [68].

Figure 10.8 shows 3D imaging results of a live rabbit heart. This system captures live rabbit heart as it beats at a rate of approximately 200 beats per minute. This specific system is composed of the TI DLP LightCrafter projector and a Vision Research Phantom V9.1 high-speed CMOS camera. The camera's resolution was 576×576 and the projector's resolution was 608×648. The projector switches binary patterns at 2000 Hz, and thus the camera was set to capture at 2000 Hz to capture image simultaneously. A two wavelength phase-shifting technique was used with the shorter wavelength being the OPWM patterns, and the longer wavelength being the Stucki dithered patterns; and a temporal phase-unwrapping algorithm was

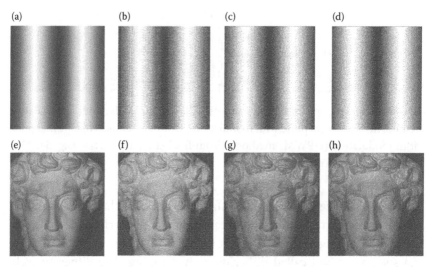

FIGURE 10.7 Binarizing grayscale images with different dithering techniques. (a) 8-Bit sinusoidal pattern. (b) Bayer-ordered dithering [188]. (c) Floyd–Steinburg error-diffusion dithering [190]. (d) Stucki error-diffusion dithering [191]. (e) 8-Bit sinusoidal pattern. (f) Bayer-ordered dithering [188]. (g) Floyd–Steinburg error-diffusion dithering [190]. (h) Stucki error-diffusion dithering [191]. (From B. Li et al., *Opt. Laser Eng.* 54, 236–246, 2014. With permission.)

used to unwrap the phase pixel by pixel. Further study showed that a speed of at least 800 Hz was required to capture the heart accurately without being distorted by motion, and thus without the employment of the above-mentioned binary defocusing techniques this would not have been possible.

Though substantial improvements were realized by adopting dithering techniques to generate binary patterns. However, because all classic binary dithering methods were developed to represent arbitrary images, they rarely consider the unique periodical structures of an image that the sinusoidal pattern embodies. As a result, directly applying those dithering algorithms may not produce the best possible quality sinusoidal structured patterns. Further optimization could lead to even better results than the dithering.

The objective of optimization is to find binary patterns that are as close as possible to the ideal sinusoidal patterns after applying Gaussian smoothing. Mathematically, we are minimizing the functional

$$\min_{B,G} \|I(x,y) - G(x,y) \otimes B(x,y)\|_F \qquad (10.1)$$

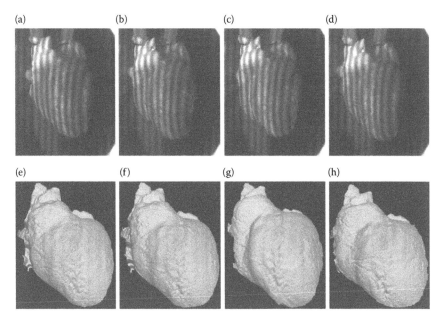

FIGURE 10.8 Example of capturing live rabbit hearts with binary defocusing techniques. (a)–(d) Captured fringes of a live rabbit heart. (e)–(h) Corresponding reconstructed 3D results of (a)–(d). (From B. Li et al., *Opt. Laser Eng.* 54, 236–246, 2014. With permission.)

where $\| \cdot \|_F$ is the Frobenius norm, $I(x, y)$ is the ideal sinusoidal intensity pattern, $G(x, y)$ is a 2D Gaussian kernel, $B(x, y)$ is the desired 2D binary pattern, and $\otimes$ represents convolution. Unfortunately, the problem is non-deterministic polynomial time (NP) hard, making it impractical to solve the problem mathematically. Furthermore, the desired pattern should perform well for different amounts of defocusing (i.e., varying $G(x, y)$), making the problem even more complex.

Various optimization strategies have been developed to further improve the quality of dithered patterns including applying a genetic algorithm [174], global phase domain optimization algorithm [192], intensity domain optimization algorithm [193], and local optimization algorithms in intensity domain [175,194].

Experiments were carried out to illustrate the differences of each different patterns. The squared binary patterns, the OPWM patterns, the Stucki dithered patterns, and the symmetry optimized patterns [175] were tested. The flat board was first measured with the fringe period of $T = 18$ pixels and the projector was nearly focused. Figure 10.9 shows the experiment

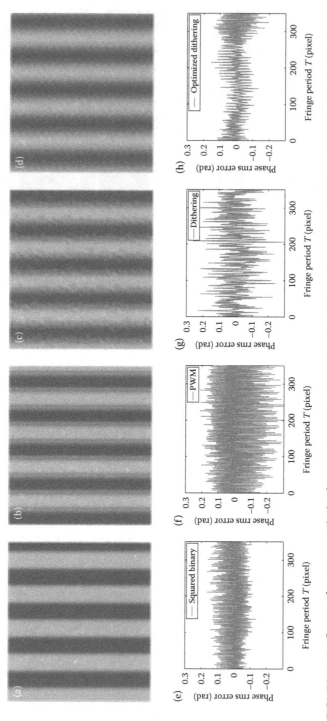

FIGURE 10.9 Captured patterns with the fringe period $T = 18$ pixels when the projector is nearly focused. (a) Squared binary pattern. (b) PWM pattern. (c) Dithering pattern. (d) Optimized-dithering pattern. (e)–(f) Corresponding error cross-section plots of (a)–(d) with standard phase rms errors of 0.08 rad, 0.15 rad, 0.10 rad, and 0.06 rad, respectively. (From B. Li et al., *Opt. Laser Eng.* 54, 236–246, 2014. With permission.)

results. It shows that optimized-dithering technique has the best performance, while OPWM technique and dithering technique fail to provide better results than SBM. Here, as well the following experiments, the phase error was computed by taking the difference between the phase obtained from these patterns with the Gaussian-smoothed phase obtained by a nine-step phase-shifting algorithm with a fringe period of $T = 18$ pixels.

The flat board was also measured using a larger fringe period (i.e., $T = 90$ pixels). Figure 10.10 shows the results when the projector was nearly focused. This experiment clearly shows that neither the SBM nor the OPWM technique could provide reasonable result, despite the little improvement that the PWM technique has over the SBM. Dithering technique and optimized-dithering technique could generate good phase, with the optimized-dithering technique being slightly better than the Stucki-dithering technique.

Finally, a 3D statue was measured to visually compare the differences among these binary techniques, as shown in Figure 10.11. The statue was measured with a fringe period of $T = 90$ pixels and with the projector being slightly defocused. The first row of the results display the captured structured images and the second row displays the rendered 3D results. The results show that when the projector is nearly in focus, the binary structure is clear for the squared binary patterns and that the optimized dithered patterns are sinusoidal in nature. It should also be noted that neither the squared binary method nor the PWM method generate high-quality 3D results, whereas the dithering and optimized-dithering techniques perform well in the sculpture's reconstruction. The last result which should be derived from this experiment is that the optimized-dithering technique is visibly better than the regular dithering method.

10.4 DISCUSSIONS

The advances in real time to superfast 3D imaging have been tremendous using the DLP platform, yet, there are still some challenging problems which should be addressed. This section presents some of these challenges we have realized but have not been able to completely overcome.

10.4.1 Challenges with Sinusoidal Fringe Projection Techniques

Even though the conventional methods of using sinusoidal fringe patterns have some challenges, their advantages are obvious: the cost is usually low and the projector is usually robust. However, to use those commercially

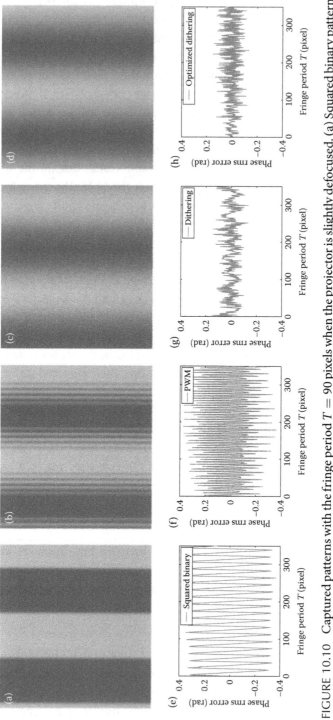

FIGURE 10.10   Captured patterns with the fringe period $T = 90$ pixels when the projector is slightly defocused. (a) Squared binary pattern. (b) PWM pattern. (c) Dithering pattern. (d) Optimized-dithering pattern. (e)–(f) Corresponding error cross-section plots of (a)–(d) with standard phase rms errors of 0.2176 rad, 0.1846 rad, 0.0534 rad, and 0.0525 rad, respectively. (From B. Li et al., *Opt. Laser Eng.* 54, 236–246, 2014. With permission.)

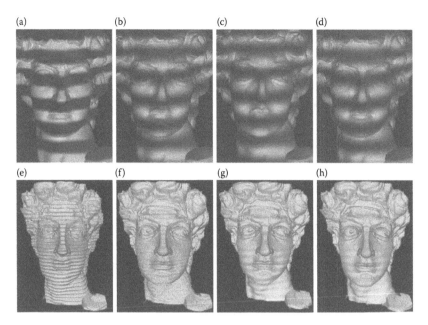

FIGURE 10.11 Complex 3D statue measurement results with the fringe period $T = 90$ pixels when the projector is slightly defocused. (a) Squared binary pattern. (b) PWM pattern. (c) Dithering pattern. (d) Optimized-dithering pattern. (e)–(h) Corresponding reconstructed 3D results of (a)–(d). (From B. Li et al., *Opt. Laser Eng.* 54, 236–246, 2014. With permission.)

available projectors for high-accuracy 3D imaging, the major challenges are as follows:

- *Nonlinear response.* Though algorithms have been developed to compensate for the error caused by the nonlinear response of the system, the nonlinear gamma effect cannot be completely eliminated. It should be noted that the nonlinear effect is not only caused by the projector's gamma, but also the camera sensor.

- *Reliability.* The majority of commercially available projectors use plastic projection lenses with focal lengths that could change due to heating. This creates huge problems for high-accuracy 3D imaging since it is very difficult to precisely calibrate the projector's parameters at its measurement condition (e.g., temperature).

- *DMD transient effect.* Wang et al. [195] found that the DMD does not respond instantaneously; this can potentially create problems for the sinusoidal method where the DMD switches ON and OFF rapidly thus making it difficult to completely eliminate this problem.

- *Synchronization requirement.* Precise synchronization is required if a DLP projection platform is used, albeit LCoS or LCD projection system does not.

- *Full dynamic range.* As demonstrated in Chapter 8, the majority of commercially available projector does not respond to the full 255 grayscale levels, and thus the dynamic range is usually not maximum.

10.4.2 Challenges with Binary Defocusing Techniques

Some speed breakthroughs have been utilizing binary defocusing techniques, yet a number of new problems arrive that affect the performance of the measurement system. The major issues associated with binary defocusing techniques are as follows:

- *Depth range.* The projector's optical engine is typically designed to generate large depth of focus, which is a great feature for conventional fringe projection methods where the projector is always nearly focused. Yet, the binary defocusing method requires the projector to be out of focus and thus the measurement range is substantially compromised. With the newly developed optimization methods on pattern generation, the depth range has be drastically increased. However, the question remains: Can we achieve the same depth range by minimally modifying the optics of a projection system?

- *High-order harmonics influence.* The binary defocusing technique essentially suppresses high-order harmonics with the defocusing effect, yet, it may not be absolute zero if the projector is nearly focused. A number of techniques have been developed to optimize the pattern such that any effects of high-order harmonics are minimized. However, we believe that there is a lot of room to further advance these methods.

- *Precise quantification of defocusing amount.* The measurement quality of the binary defocusing method is highly dependent upon the defocusing amount: too much or too little defocusing will result in a lower quality of measurement. The current practice heavily relies on experience; precisely and quantitatively measuring the level of defocusing such that the highest measurement quality can be achieved every time is still an open research problem for us.

- *Best dithered pattern.* The dithering and its optimization methods proved extremely successful to advance the binary defocusing method, but none of those optimization algorithms can ensure the global solution. It seems natural to perform an exhaustive search (e.g., try all possible combinations). However, an exhaustive searching method is not practical even for the local optimization methods because it involves extremely intensive computations: for example, to find the best 30×10 pixel pattern under one amount of defocusing (i.e., one blurring filter size), it takes $2^{30 \times 10}$ trials. The question remains: Is there a mathematical solution to this optimization problem?

10.4.3 Application Challenges

Real time to superfast 3D imaging techniques cannot thrive and continue to mature without solving real-world problems. There are some practical issues that challenge the majority of current real-time (even static) optical 3D imaging techniques, however, including

- *Automatic exposure.* Since autoexposure is very common for 2D imaging techniques, one assume this can be done easily for 3D imaging techniques as well. However important this feature is, given that a projector is involved and given that this projector is a digital device, some nontrivial issues arise.

- *Automatic focus.* Again, autofocus is a critical function of the majority 2D imaging systems, can we do autofocusing for 3D imaging system? If we can, how can we calibrate the system after the focal length of the system is changed?

- *Shiny part measurement.* Optical methods suffer if the sensor sees a saturated image or an image without enough intensity. Shiny parts create both problems due to very high surface contrast. We believe such a problem could be at least partially solved through the usage of high dynamic range (HDR) methods as used in 2D techniques, yet it remains an open problem on how to rapidly and automatically achieve HDR for 3D techniques.

- *Harsh industrial environment usage.* One of the largest application areas for 3D imaging is quality control in the manufacturing field. However, the real-world environment is not always pleasant: high temperatures, high vibrations, flash lighting, etc. and all these pose challenges for adopting 3D optical imaging systems for on-site applications.

- *Large outdoor applications.* Yet another application of a superfast 3D imaging system is to capture dynamics of real scenes; this might involve a system to capture both inside and outside of a room. Since this technique is optical base, sunlight could pose challenges.

- *Big data storage.* 3D imaging data are much larger than its 2D counterparts. For example, to store a 45-s 3D video with standard OBJ format, it will use approximately 42 GB if the video is captured at a resolution of 640 (fairly low), and 30 Hz [196]. Compression methods are needed to naturally compress recorded 3D data, like what have been done in 2D imaging technology. Even more challenging problem is to compress 3D data while it is recording.

- *Big data analytics.* 3D data will not be useful unless we can make sense of the data. 3D data are naturally big data problem that requires a lot investigations to analyze them and utilize them for applications. For example, to perform quality control, 3D captured data must be compared with the ideal one (e.g., a CAD design) to generate an error map, which might not be straightforward due to noise and other influences.

10.5 SUMMARY REMARKS

This chapter has summarized the route to achieve real time to kHz 3D imaging speeds by leveraging the DLP technologies that were invented by TI. These technologies have already seen extensive applications, and there are many more applications in which these technologies can be applied. The fundamental barriers of achieving superfast 3D imaging were successfully overcome through different innovations, but there are still a number of challenges existing. We believe further innovations are required to allow real time to superfast imaging technologies to be adopted on a much larger scale.

This book has covered a variety of topics that is important to successfully develop 3D imaging systems, explained the principles of each technology, and provided an example of developing a practical 3D imaging system. This book only presented our own perspective on 3D imaging system development with DFP techniques. It is important to note that this book, by no means, is to argue that the approaches or methods presented in this book are the best available. The reader is encouraged to use his/her own judgements to decide the best practical solutions for their particular needs.

References

1. S. Zhang, ed., *Handbook of 3D Machine Vision: Optical Metrology and Imaging*, 1st ed. (CRC Press, New York, NY), 2013.
2. A. Kolb, E. Barth, and R. Koch, Time-of-flight cameras in computer graphics, *Comput. Graph. Forum* 29(1), 141–159, 2010.
3. C. Filiberto, C. Roberto, P. Dario, and R. Fulvio, Sensors for 3D imaging: Metric evaluation and calibration of a CCD/CMOS time-of-flight camera, *Sensors* 9(12), 10,080–10,096, 2009.
4. S. Foix, G. Alenya, and C. Torras, Lock-in time-of-flight (ToF) cameras: A survey, *IEEE Sens. J.* 11(9), 1917–1926, 2011.
5. C. P. Keferstein and M. Marxer, Testing bench for laser triangulation sensors, *Sensor Rev.* 18(3), 183–187, 1998.
6. A. Sirikasemlert and X. Tao, Objective evaluation of textural changes in knitted fabrics by laser triangulation, *Text. Res. J.* 70(12), 1076–1087, 2000.
7. G. Manneberg, S. Hertegard, and J. Liljencrantz, Measurement of human vocal fold vibrations with laser triangulation, *Opt. Eng.* 40(9), 2041–2044, 2001.
8. U. R. Dhond and J. K. Aggarwal, Structure from stereo—A review, *IEEE Trans. Syst. Man Cybern.* 19(6), 1489–1510, 1989.
9. D. Scharstein and R. Szeliski, A taxonomy and evaluation of dense two-frame stereo correspondence algorithms, *Int. J. Comp. Vis.* 47(1–3), 7–42, 2002.
10. R. I. Hartley and A. Zisserman, *Multiple View Geometry in Computer Vision* (Cambridge University Press, ISBN: 0521623049), 2000.
11. T. Kanade and M. Okutomi, A stereo matching algorithm with an adaptive window: Theory and experiment, *IEEE Trans. Pattern Anal. Mach. Itell.* 16(9), 920–932, 1994.
12. V. Kolmogorov and R. Zabih, Multi-camera scene reconstruction via graph cuts, *European Conference on Computer Vision*, 82–96, (Copenhagen, Denmark), May 28–31, 2002.
13. J. Kostková and R. Sára, Stratified dense matching for stereopsis in complex scenes. *Proceedings of British Machine Vision Conference*, 339–348, (Norfolk, United Kingdom), September 09–11, 2003.
14. H. Hirschmuller, Stereo processing by semiglobal matching and mutual information, *IEEE Trans. Pattern Anal. Mach. Itell.* 30(2), 328–341, 2008.
15. F. Besse, C. Rother, A. W. Fitzgibbon, and J. Kautz, PMBP: PatchMatch belief propagation for correspondence field estimation, *Int. J. Comp. Vis.* 110(1), 2–13, 2013.
16. A. Geiger, M. Roser, and R. Urtasun, Efficient large-scale stereo matching, *Lecture Notes in Computer Science*. 6492, 25–38, 2011.
17. J. Salvi, S. Fernandez, T. Pribanic, and X. Llado, A state of the art in structured light patterns for surface profilometry, *Pattern Recogn.* 43(8), 2666–2680, 2010.

18. Y. Huang, Y. Shang, Y. Liu, and H. Bao, *Handbook of 3D Machine Vision: Optical Metrology and Imaging*, chap. 3D Shapes from Speckle, 33–56, 1st ed. (CRC, 2013).
19. J. Geng, Structured-light 3D surface imaging: A tutorial, *Adv. Opt. Photonics* 3(2), 128–160, 2011.
20. Z. Zhang, Microsoft Kinect sensor and its effect, *IEEE Multimedia* 19(2), 4–10, 2012.
21. S. Zhang, Recent progresses on real-time 3-D shape measurement using digital fringe projection techniques, *Opt. Laser Eng.* 48(2), 149–158, 2010.
22. J. Pan, P. Huang, S. Zhang, and F.-P. Chiang, Color n-ary gray code for 3-D shape measurement, *12th International Conference on Experimental Mechanics* (Politecnico di Bari, Italy), 2004.
23. B. Carrihill and R. Hummel, Experiments with the intensity ratio depth sensor, *Comp. Vision Graph. Image Process.* 32, 337–358, 1985.
24. G. Chazan and N. Kiryati, *Pyramidal Intensity-Ratio Depth Sensor*, Tech. Rep., Israel Institute of Technology, Technion, Haifa, Israel, 1995.
25. P. Jia, J. Kofman, and C. English, Two-step triangular-pattern phase-shifting method for three-dimensional object-shape measurement, *Opt. Eng.* 46(8), 083201, 2007.
26. P. S. Huang, S. Zhang, and F.-P. Chiang, Trapezoidal phase-shifting method for three-dimensional shape measurement, *Opt. Eng.* 44(12), 123601, 2005.
27. M. Takeda and K. Mutoh, Fourier transform profilometry for the automatic measurement of 3-D object shapes, *Appl. Opt.* 22, 3977–3982, 1983.
28. X. Su and Q. Zhang, Dynamic 3-D shape measurement method: A review, *Opt. Laser. Eng.* 48, 191–204, 2010.
29. Q. Kemao, Windowed Fourier transform for fringe pattern analysis, *Appl. Opt.* 43, 2695–2702, 2004.
30. H. Guo and P. Huang, 3-D shape measurement by use of a modified Fourier transform method, *Proceedings of SPIE*, 7066, 70660E, SPIE Optics and Photonics (San Diego, California), August 10–14, 2008.
31. J. Li, X.-Y. Su, and L.-R. Guo, Improved Fourier transform profilometry for the automatic measurement of three-dimensional object shapes, *Opt. Eng.* 29(12), 1439–1444, 1990.
32. H. Schreiber and J. H. Bruning, *Optical Shop Testing*, chap. Phase shifting interferometry, 547–655, 3rd ed. (John Wiley & Sons, New York, NY), 2007.
33. M. Subbarao and G. Surya, Depth from defocus: A spatial domain approach, *Int. J. Comput. Vision* 13, 271–294, 1994.
34. V. Aslantas, A depth estimation algorithm with a single image, *Opt. Express* 15, 5024–5029, 2007.
35. J. Gibson, *The Perception of the Visual World* (Greenwood Pub Group, Westport, Connecticut), 1950.
36. J. Garding, Surface orientation and curvature from differential texture distortion, *Comput. Vision* 20, 733–739, 1995.
37. M. W. Pettet, Shape and contour detection, *Vision Res.* 39, 551–557, 1999.
38. J. Malik and R. Rosenholtz, Computing local surface orientation and shape from texture for curved surfaces, *Int. J. Comput. Vision* 23, 149–168, 1997.
39. T. Zhang and I. Yamaguchi, Three-dimensional microscopy with phase-shifting digital holography, *Opt. Lett.* 23(15), 1221–1223, 1998.
40. G. Pedrini, P. Froning, H. Fessler, and H. J. Tiziani, In-line digital holographic interferometry, *Appl. Opt.* 37(26), 6262–6269, 1998.
41. G. Pedrini, P. Froning, H. J. Tiziani, and M. E. Gusev, Pulsed digital holography for high-speed contouring that uses a two-wavelength method, *Appl. Opt.* 38(16), 3460–3467, 1999.

42. G. Pan and H. Meng, Digital holography of particle fields: Reconstruction by use of complex amplitude, *Appl. Opt.* 42(5), 827–833, 2003.

43. T. Bell, N. Karpinsky, and S. Zhang, *High-Resolution, Real-Time 3D Sensing with Structured Light Techniques*, chap. 4, Interactive Displays (John Wiley & Sons), 2014. (invited).

44. L. Ekstrand, Y. Wang, N. Karpinsky, and S. Zhang, Superfast 3D profilometry with digital fringe projection and phase-shifting techniques, in Song Zhang, ed., *Handbook of 3D Machine Vision: Optical Metrology and Imaging* (CRC Taylor & Francis Group, Chapter 9, 233–251), 2013 (invited).

45. M. Takeda, Fourier fringe analysis and its applications to metrology of extreme physical phenomena: A review, *Appl. Opt.* 52(1), 20–29, 2013.

46. D. C. Ghiglia and M. D. Pritt, eds., *Two-Dimensional Phase Unwrapping: Theory, Algorithms, and Software* (John Wiley & Sons, New York), 1998.

47. S. Zhang and P. S. Huang, Novel method for structured light system calibration, *Opt. Eng.* 45(8), 083601, 2006.

48. Q. Kemao, Two-dimensional windowed Fourier transform for fringe pattern analysis: Principles, applications and implementations, *Opt. Laser. Eng.* 45, 304–317, 2007.

49. D. Malacara, ed., *Optical Shop Testing*, 3rd ed. (John Wiley & Sons, New York, NY), 2007.

50. K. Creath, Phase-measurement interferometry techniques, *Progress in Optics* 26(26), 349–393.

51. L. Ekstrand and S. Zhang, Three-dimensional profilometry with nearly focused binary phase-shifting algorithms, *Opt. Lett.* 36(23), 4518–4520, 2011.

52. J. E. Greivenkamp, Generalized data reduction for heterodyne interferometry, *Opt. Eng.* 23(4), 350–352, 1984.

53. J. H. Brunning, D. R. Herriott, J. E. Gallagher, D. P. Rosenfeld, A. D. White, and D. J. Brangaccio, Digital wavefront measuring interferometer for testing optical surfaces, lenses, *Appl. Opt.* 13(11), 2693–2703, 1974.

54. P. Carré, Installation et utilisation du comparateur photoélectrigue et interferentiel du bureau international des Poids ek Measures, *Metrologia* 2(1), 13–23, 1966.

55. P. Hariharan, B. F. Oreb, and T. Eiju, Digital phase-shifting interferometry: A simple error-compensation phase calculation algorithm, *Appl. Opt.* 26(13), 2504–2506, 1987.

56. Y.-Y. Cheng and J. C. Wyant, Two-wavelength phase shifting interferometry, *Appl. Opt.* 23, 4539–4543, 1984.

57. Y.-Y. Cheng and J. C. Wyant, Multiple-wavelength phase shifting interferometry, *Appl. Opt.* 24, 804–807, 1985.

58. K. Creath, Step height measurement using two-wavelength phase-shifting interferometry, *Appl. Opt.* 26, 2810–2816, 1987.

59. D. P. Towers, J. D. C. Jones, and C. E. Towers, Optimum frequency selection in multi-frequency interferometry, *Opt. Lett.* 28, 1–3, 2003.

60. L. J. Hornbeck, Digital light processing for high-brightness high-resolution applications, *Proceedings of SPIE* 3013, 27–40, SPIE Photonics West (San Jose, California), February 10–12, 1997.

61. B. Li, Y. Wang, J. Dai, W. Lohry, and S. Zhang, Some recent advances on superfast 3D shape measurement with digital binary defocusing techniques, *Opt. Laser Eng.* 54, 236–246, 2014.

62. S. Zhang and P. S. Huang, High-resolution real-time three-dimensional shape measurement, *Opt. Eng.* 45(12), 123601, 2006.

63. B. Li, J. Gibson, J. Middendorf, Y. Wang, and S. Zhang, Comparison between LCOS projector and DLP projector in generating digital sinusoidal fringe patterns, SPIE Optics and Photonics (San Diego, California), 2013.

64. S. Lei and S. Zhang, Flexible 3-D shape measurement using projector defocusing, *Opt. Lett.* 34(20), 3080–3082, 2009.

65. S. Lei and S. Zhang, Digital sinusoidal fringe generation: Defocusing binary patterns VS focusing sinusoidal patterns, *Opt. Laser Eng.* 48(5), 561–569, 2010.

66. Y. Wang and S. Zhang, Comparison among square binary, sinusoidal pulse width modulation, and optimal pulse width modulation methods for three-dimensional shape measurement, *Appl. Opt.* 51(7), 861–872, 2012.

67. Y. Xu, L. Ekstrand, J. Dai, and S. Zhang, Phase error compensation for three-dimensional shape measurement with projector defocusing, *Appl. Opt.* 50(17), 2572–2581, 2011.

68. Y. Wang, J. I. Laughner, I. R. Efimov, and S. Zhang, 3D absolute shape measurement of live rabbit hearts with a superfast two-frequency phase-shifting technique, *Opt. Express* 21(5), 5822–5632, 2013.

69. N. Karpinsky, M. Hoke, V. Chen, and S. Zhang, High-resolution, real-time three-dimensional shape measurement on graphics processing unit, *Opt. Eng.* 53(2), 024105, 2014.

70. K. Liu, Y. Wang, D. L. Lau, Q. Hao, and L. G. Hassebrook, Dual-frequency pattern scheme for high-speed 3-D shape measurement, *Opt. Express* 18, 5229–5244, 2010.

71. G. Sansoni, M. Carocci, and R. Rodella, Three-dimensional vision based on a combination of gray-code and phase-shift light projection: Analysis and compensation of the systematic errors, *Appl. Opt.* 38, 6565–6573, 1999.

72. S. Zhang, Flexible 3D shape measurement using projector defocusing: Extended measurement range, *Opt. Lett.* 35(7), 931–933, 2010.

73. O. Hall-Holt and S. Rusinkiewicz, Stripe boundary codes for real-time structured-light range scanning of moving objects, *The 8th IEEE International Conference on Computer Vision*, II: 359–366, (Vancouver, Canada), July 7–14, 2001.

74. Y. Wang, S. Zhang, and J. H. Oliver, 3-D shape measurement technique for multiple rapidly moving objects, *Opt. Express* 19(9), 5149–5155, 2011.

75. Q. Z. X. Su, L. Xiang, and X. Sun, 3-D shape measurement based on complementary gray-code light, *Opt. Laser Eng.* 50, 574–579, 2012.

76. J. Davis, R. Ramamoorthi, and S. Rusinkiewicz, Spacetime stereo: A unifying framework for depth from triangulation, *2003 IEEE Computer Society Conference on Computer Vision and Pattern Recognition*, 2, II-359-66, (Toronto, Canada), June 16–22, 2003.

77. Y. Wang and S. Zhang, Novel phase coding method for absolute phase retrieval, *Opt. Lett.* 37(11), 2067–2069, 2012.

78. C. Zhou, T. Liu, S. Si, J. Xu, Y. Liu, and Z. Lei, Phase coding method for absolute phase retrieval with a large number of codewords, *Opt. Express* 20(22), 24,139–24,150, 2012.

79. C. Zhou, T. Liu, S. Si, J. Xu, Y. Liu, and Z. Lei, An improved stair phase encoding method for absolute phase retrieval, *Opt. Laser Eng.* 66, 269–278, 2015.

80. S. Zhang, Composite phase-shifting algorithm for absolute phase measurement, *Opt. Laser Eng.* 50(11), 1538–1541, 2012.

81. N. Karpinsky and S. Zhang, Composite phase-shifting algorithm for three-dimensional shape compression, *Opt. Eng.* 49(6), 063604, 2010.

82. Y. Li, C. Zhao, Y. Qian, H. Wang, and H. Jin, High-speed and dense three-dimensional surface acquisition using defocused binary patterns for spatially isolated objects, *Opt. Express* 18(21), 21,628–21,635, 2010.

83. Y. Li, H. Jin, and H. Wang, Three-dimensional shape measurement using binary spatio-temporal encoded illumination, *J. Opt. A: Pure Appl. Opt.* 11(7), 075502, 2009.

84. A. Wiegmann, H. Wagner, and R. Kowarschik, Human face measurement by projecting bandlimited random patterns, *Opt. Express* 14(17), 7692–7698, 2006.

85. L. Zhang, N. Snavely, B. Curless, and S. M. Seitz, Spacetime faces: High-resolution capture for modeling and animation, *ACM Trans. Graph.* 23(3), 548–558, 2004.

86. L. Zhang, B. Curless, and S. Seitz, Spacetime stereo: Shape recovery for dynamic scenes, *Proceedings of the Computer Vision and Pattern Recognition*, 367–374, (Toronto, Canada), June 16–22, 2003.

87. W. Jang, C. Je, Y. Seo, and S. W. Lee, Structured-light stereo: Comparative analysis and integration of structured-light and active stereo for measuring dynamic shape, *Opt. Laser Eng.* 51(11), 1255–1264, 2013.

88. M. Schaffer, M. Große, B. Harendt, and R. Kowarschik, Coherent two-beam interference fringe projection for high-speed three-dimensional shape measurements, *Appl. Opt.* 52(11), 2306–2311, 2013.

89. K. Liu and Y. Wang, Phase channel multiplexing pattern strategy for active stereo vision, *International Conference on 3D Imaging (IC3D)*, 1–8, (Liege, Belgium), December 3–5, 2012.

90. Z. Li, K. Zhong, Y. Li, X. Zhou, and Y. Shi, Multiview phase shifting: A full-resolution and high-speed 3D measurement framework for arbitrary shape dynamic objects, *Opt. Lett.* 38(9), 1389–1391, 2013.

91. K. Zhong, Z. Li, Y. Shi, C. Wang, and Y. Lei, Fast phase measurement profilometry for arbitrary shape objects without phase unwrapping, *Opt. Laser Eng.* 51(11), 1213–1222, 2013.

92. J. Pan, P. S. Huang, and F.-P. Chiang, Color phase-shifting technique for three-dimensional shape measurement, *Opt. Eng.* 45(12), 013602, 2006.

93. W. Lohry, V. Chen, and S. Zhang, Absolute three-dimensional shape measurement using coded fringe patterns without phase unwrapping or projector calibration, *Opt. Express* 22(2), 1287–1301, 2014.

94. M. Maruyama and S. Abe, Range sensing by projecting multiple slits with random cuts, *IEEE Trans. Pattern Anal. Mach. Intell.* 15(6), 647–651, 1993.

95. K. Konolige, Projected texture stereo, *IEEE International Conference on Robotics and Automation*, 148–155, (Anchorage, Alaska), May 3–7, 2010.

96. W. Lohry and S. Zhang, High-speed absolute three-dimensional shape measurement using three binary dithered patterns, *Opt. Express* 22(22), 26,752–26,762, 2014.

97. S. Zhang, X. Li, and S.-T. Yau, Multilevel quality-guided phase unwrapping algorithm for real-time three-dimensional shape reconstruction, *Appl. Opt.* 46(1), 50–57, 2007.

98. W. Gao and K. Qian, Parallel computing in experimental mechanics and optical measurement: A review, *Opt. Laser Eng.* 50(4), 608–617, 2012.

99. J. M. Huntley, Noise-immune phase unwrapping algorithm, *Appl. Opt.* 28, 3268–3270, 1989.

100. R. M. Goldstein, H. A. Zebker, and C. L. Werner, Two-dimensional phase unwrapping, *Radio Sci.* 23, 713–720, 1988.

101. R. Cusack, J. M. Huntley, and H. T. Goldrein, Improved noise-immune phase unwrapping algorithm, *Appl. Opt.* 34, 781–789, 1995.

102. J. R. Buchland, J. M. Huntley, and S. R. E. Turner, Unwrapping noisy phase maps by use of a minimum-cost-matching algorithm, *Appl. Opt.* 34, 5100–5108, 1995.

103. M. F. Salfity, P. D. Ruiz, J. M. Huntley, M. J. Graves, R. Cusack, and D. A. Beauregard, Branch cut surface placement for unwrapping of undersampled three-dimensional phase data: Application to magnetic resonance imaging arterial flow mapping, *Appl. Opt.* 45, 2711–2722, 2006.

104. T. J. Flynn, Two-dimensional phase unwrapping with minimum weighted discontinuity, *J. Opt. Soc. Am. A* 14, 2692–2701, 1997.

105. D. C. Ghiglia and L. A. Romero, Minimum L^p-norm two-dimensional phase unwrapping, *J. Opt. Soc. Am. A* 13, 1–15, 1996.

106. A. Baldi, Phase unwrapping by region growing, *Appl. Opt.* 42, 2498–2505, 2003.

107. K. M. Hung and T. Yamada, Phase unwrapping by regions using least-squares approach, *Opt. Eng.* 37, 2965–2970, 1998.

108. M. A. Merráez, J. G. Boticario, M. J. Labor, and D. R. Burton, Agglomerative clstering-based approach for two-dimensional phase unwrapping, *Appl. Opt.* 44, 1129–1140, 2005.

109. J.-J. Chyou, S.-J. Chen, and Y.-K. Chen, Two-dimensional phase unwrapping with a multichannel least-mean-square algorithm, *Appl. Opt.* 43, 5655–5661, 2004.

110. X. Su and W. Chen, Reliability-guided phase unwrapping algorithm: A review, *Opt. Laser Eng.* 42(3), 245–261, 2004.

111. M. Zhao, L. Huang, Q. Zhang, X. Su, A. Asundi, and Q. Kemao, Quality-guided phase unwrapping technique: Comparison of quality maps and guiding strategies, *Appl. Opt.* 50(33), 6214–6224, 2011.

112. D. J. Bone, Fourier fringe analysis: The two-dimensional phase unwrapping problem, *Appl. Opt.* 30, 3627–3632, 1991.

113. J. A. Quiroga, A. Gonzalez-Cano, and E. Bernabeu, Phase-unwrapping algorithm based on an adaptive criterion, *Appl. Opt.* 34, 2560–2563, 1995.

114. M. D. Pritt, Phase-unwrapping by means of multigrid techniques for interferometric SAR, *IEEE Trans. Geosci. Remote Sens.* 34, 728–738, 1996.

115. B. Ströbel, Processing of interferometric phase maps as complex-valued phasor images, *Appl. Opt.* 35, 2192–2198, 1996.

116. J.-L. Li, X.-Y. Su, and J.-T. Li, Phase unwrapping algorithm based on reliability and edge detection, *Opt. Eng.* 36, 1685–1690, 1997.

117. M. A. Herráez, D. R. Burton, M. J. Lalor, and M. A. Gdeisat, Fast two-dimensional phase-unwrapping algorithm based on sorting by reliability following a noncontinuous path, *Appl. Opt.* 41, 7437–7444, 2002.

118. C. Quan, C. J. Tay, L. Chen, and Y. Fu, Spatial-fringe-modulation-based quality map for phase unwrapping, *Appl. Opt.* 42, 7060–7065, 2003.

119. S. Zhang and S.-T. Yau, High-resolution, real-time 3-D absolute coordinate measurement based on a phase-shifting method, *Opt. Express* 14(7), 2644–2649, 2006.

120. P. S. Huang and S. Zhang, A fast three-step phase-shifting algorithm, *Proceedings of SPIE*, 6000, 60,000F (Boston, MA), 2005.

121. P. S. Huang, Q. J. Hu, and F.-P. Chiang, Double three-step phase-shifting algorithm, *Appl. Opt.* 41(22), 4503–4509, 2002.

122. S. Zhang, High-resolution 3D profilometry with binary phase-shifting methods, *Appl. Opt.* 50(12), 1753–1757, 2011.

123. X. Y. Su, W. S. Zhou, G. Von Bally, and D. Vukicevic, Automated phase-measuring profilometry using defocused projection of a Ronchi grating, *Opt. Commun.* 94, 561–573, 1992.

124. P. S. Huang, C. Zhang, and F.-P. Chiang, High-speed 3-D shape measurement based on digital fringe projection, *Opt. Eng.* 42, 163–168, 2003.

125. X. Zhang, L. Zhu, Y. Li, and D. Tu, Generic nonsinusoidal fringe model and gamma calibration in phase measuring profilometry, *J. Opt. Soc. Am. A* 29(6), 1047–1058, 2012.

126. B. Pan, Q. Kemao, L. Huang, and A. Asundi, Phase error analysis and compensation for nonsinusoidal waveforms in phase-shifting digital fringe projection profilometry, *Opt. Lett.* 34(4), 2906–2914, 2009.

127. S. Zhang and P. S. Huang, Phase error compensation for a three-dimensional shape measurement system based on the phase shifting method, *Opt. Eng.* 46(6), 063601, 2007.

128. S. Zhang and S.-T. Yau, Generic nonsinusoidal phase error correction for three-dimensional shape measurement using a digital video projector, *Appl. Opt.* 46(1), 36–43, 2007.

129. H. Guo, H. He, and M. Chen, Gamma correction for digital fringe projection profilometry, *Appl. Opt.* 43, 2906–2914, 2004.

130. K. Liu, Y. Wang, D. L. Lau, Q. Hao, and L. G. Hassebrook, Gamma model and its analysis for phase measuring profilometry, *J. Opt. Soc. Am. A* 27(3), 553–562, 2010.

131. T. Hoang, B. Pan, B. Nguyen, and Z. Wang, Generic gamma correction for accuracy enhancement in fringe projection profilometry, *Opt. Lett.* 35(2), 1992–1994, 2010.

132. Z. Li and Y. Li, Gamma-distorted fringe image modeling and accurate gamma correction for fast phase measuring profilometry, *Opt. Lett.* 36(2), 154–156, 2011.

133. P. Zhou, X. Liu, Y. He, and T. Zhu, Phase error analysis and compensation considering ambient light for phase measuring profilometry, *Opt. Laser Eng.* 55, 99–104, 2014.

134. S. Ma, R. Zhu, C. Quan, B. Li, C. J. Tay, and L. Chen, Blind phase error suppression for color-encoded digital fringe projection profilometry, *Opt. Commun.* 285, 1662–1668, 2012.

135. S. Ma, C. Quan, R. Zhu, L. Chen, B. Li, and C. J. Tay, A fast and accurate gamma correction based on Fourier spectrum analysis for digital fringe projection profilometry, *Opt. Laser Eng.* 285, 533–538, 2012.

136. S. Zhang, Comparative study on passive and active projector nonlinear gamma calibration, *Appl. Opt.* 54(13), 3834–3841, 2015.

137. B. Li, N. Karpinsky, and S. Zhang, Novel calibration method for structured light system with an out-of-focus projector, *Appl. Opt.* 56(13), 3415–3426, 2014.

138. Y. Wen, S. Li, H. Cheng, X. Su, and Q. Zhang, Universal calculation formula and calibration method in Fourier transform profilometry, *Appl. Opt.* 49(34), 6563–6569, 2010.

139. Y. Xiao, Y. Cao, and Y. Wu, Improved algorithm for phase-to-height mapping in phase measuring profilometry, *Appl. Opt.* 51(8), 1149–1155, 2012.

140. Y. Villa, M. Araiza, D. Alaniz, R. Ivanov, and M. Ortiz, Transformation of phase to (x,y,z)-coordinates for the calibration of a fringe projection profilometer, *Opt. Laser Eng.* 50(2), 256–261, 2012.

141. C. B. Duane, Close-range camera calibration, *Photogramm. Eng.* 37(8), 855–866, 1971.

142. I. Sobel, On calibrating computer controlled cameras for perceiving 3-D scenes, *Artif. Intell.* 5(2), 185–198, 1974.

143. R. Tsai, A versatile camera calibration technique for high-accuracy 3D machine vision metrology using off-the-shelf TV cameras and lenses, *IEEE J. Robot. Auto.* 3(4), 323–344, 1987.

144. Z. Zhang, A flexible new technique for camera calibration, *IEEE Trans. Pattern Anal. Mach. Intell.* 22(11), 1330–1334, 2000.

145. J. Lavest, M. Viala, and M. Dhome, Do we really need an accurate calibration pattern to achieve a reliable camera calibration? *European Conference on Computer Vision*, 158–174, (Freiburg, Germany), June 2–6, 1998.

146. A. Albarelli, E. Rodolà, and A. Torsello, Robust camera calibration using inaccurate targets, *Trans. Pattern Anal. Mach. Intell.* 31(2), 376–383, 2009.

147. K. H. Strobl and G. Hirzinger, More accurate pinhole camera calibration with imperfect planar target, *IEEE International Conference on Computer Vision Workshops (ICCV Workshops)*, 1068–1075, (Barcelona, Spain), November 6–13, 2011.

148. L. Huang, Q. Zhang, and A. Asundi, Flexible camera calibration using not-measured imperfect target, *Appl. Opt.* 52(25), 6278–6286, 2013.

149. C. Schmalz, F. Forster, and E. Angelopoulou, Camera calibration: Active versus passive targets, *Opt. Eng.* 50(11), 113,601–113,601, 2011.

150. L. Huang, Q. Zhang, and A. Asundi, Camera calibration with active phase target: Improvement on feature detection and optimization, *Opt. Lett.* 38(9), 1446–1448, 2013.

151. Q. Hu, P. S. Huang, Q. Fu, and F.-P. Chiang, Calibration of a three-dimensional shape measurement system, *Opt. Eng.* 42(2), 487–493, 2003.

152. X. Mao, W. Chen, and X. Su, Improved Fourier-transform profilometry, *Appl. Opt.* 46(5), 664–668, 2007.

153. E. Zappa and G. Busca, Fourier-transform profilometry calibration based on an exhaustive geometric model of the system, *Opt. Laser Eng.* 47(7), 754–767, 2009.

154. H. Guo, M. Chen, and P. Zheng, Least-squares fitting of carrier phase distribution by using a rational function in fringe projection profilometry, *Opt. Lett.* 31(24), 3588–3590, 2006.

155. H. Du and Z. Wang, Three-dimensional shape measurement with an arbitrarily arranged fringe projection profilometry system, *Opt. Lett.* 32(16), 2438–2440, 2007.

156. L. Huang, P. S. Chua, and A. Asundi, Least-squares calibration method for fringe projection profilometry considering camera lens distortion, *Appl. Opt.* 49(9), 1539–1548, 2010.

157. M. Vo, Z. Wang, B. Pan, and T. Pan, Hyper-accurate flexible calibration technique for fringe-projection-based three-dimensional imaging, *Opt. Express* 20(15), 16,926–16,941, 2012.

158. R. Legarda-Sáenz, T. Bothe, and W. P. Ju, Accurate procedure for the calibration of a structured light system, *Opt. Eng.* 43(2), 464–471, 2004.

159. Z. Li, Y. Shi, C. Wang, and Y. Wang, Accurate calibration method for a structured light system, *Opt. Eng.* 47(5), 053, 604, 2008.

160. Y. Yin, X. Peng, A. Li, X. Liu, and B. Z. Gao, Calibration of fringe projection profilometry with bundle adjustment strategy, *Opt. Lett.* 37(4), 542–544, 2012.

161. D. Han, A. Chimienti, and G. Menga, Improving calibration accuracy of structured light systems using plane-based residual error compensation, *Opt. Eng.* 52(10), 104, 106, 2013.

162. L. Merner, Y. Wang, and S. Zhang, Accurate calibration for 3D shape measurement system using a binary defocusing technique, *Opt. Laser Eng.* 51(5), 514–519, 2013.

163. J.-Y. Bouguet, Camera Calibration ToolBox for Matlab, Online available at: http://www.vision.caltech.edu/bouguetj/calib_doc.

164. S. Zhang, D. Royer, and S. T. Yau, GPU-assisted high-resolution, real-time 3-D shape measurement, *Opt. Express* 14(20), 9120–9129, 2006.

165. B. Li and S. Zhang, Structured light system calibration method with optimal fringe angle, *Appl. Opt.* 53(33), 7942–7950, 2014.
166. Y. Wang and S. Zhang, Optimal fringe angle selection for digital fringe projection technique, *Appl. Opt.* 52(29), 7094–7098, 2013.
167. W. Lohry, Y. Xu, and S. Zhang, Optimal checkerboard selection for structured light system calibration, *Proceedings of SPIE*, 7432, 743202 (San Diego, CA), 2009.
168. S. Gorthi and P. Rastogi, Fringe projection techniques: Whither we are? *Opt. Laser Eng.* 48, 133–140, 2010.
169. S. Hsu, S. Acharya, A. Rafii, and R. New, Performance of a time-of-flight range camera for intelligent vehicle safety applications, in J. Vallldorf, and W. Gessner eds., *Advanced Microsystems for Automotive Applications* (Springer Berlin Heidelberg), 205–219, 2006.
170. S. Rusinkiewicz, O. Hall-Holt, and M. Levoy, Real-time 3D model acquisition, *ACM Trans. Graph.* 21(3), 438–446, 2002.
171. S. Zhang, D. van der Weide, and J. Oliver, Superfast phase-shifting method for 3-D shape measurement, *Opt. Express* 18(9), 9684–9689, 2010.
172. Y. Wang and S. Zhang, Optimal pulse width modulation for sinusoidal fringe generation with projector defocusing, *Opt. Lett.* 35(24), 4121–4123, 2010.
173. Y. Wang and S. Zhang, Three-dimensional shape measurement with binary dithered patterns, *Appl. Opt.* 51(27), 6631–6636, 2012.
174. W. Lohry and S. Zhang, Genetic method to optimize binary dithering technique for high-quality fringe generation, *Opt. Lett.* 38(4), 540–542, 2013.
175. J. Dai, B. Li, and S. Zhang, High-quality fringe patterns generation using binary pattern optimization through symmetry and periodicity, *Opt. Laser Eng.* 52, 195–200, 2014.
176. P. S. Huang and S. Zhang, Fast three-step phase-shifting algorithm, *Appl. Opt.* 45(21), 5086–5091, 2006.
177. S. Zhang and P. Huang, High-resolution, real-time 3-D shape acquisition, *IEEE Computer Vision and Pattern Recognition*, 3, 28–37 (Washington DC, MD), 2004.
178. Y. Wang and S. Zhang, Superfast multifrequency phase-shifting technique with optimal pulse width modulation, *Opt. Express* 19(6), 5143–5148, 2011.
179. S. Zhang, Y. Wang, J. I. Laughner, and I. R. Efimov, Measuring dynamic 3D microstructures using a superfast digital binary phase-shifting technique, *ASME 2013 International Manufacturing Science and Engineering Conference* (Madison, Wisconsin), June 10–14, 2013.
180. H. Fujita, K. Yamatan, M. Yamamoto, Y. Otani, A. Suguro, S. Morokawa, and T. Yoshizawa, Three-dimensional profilometry using liquid crystal grating, *Proceedings of SPIE*, 5058, 51–60 (Beijing, China), 2003.
181. T. Yoshizawa and H. Fujita, Liquid crystal grating for profilometry using structured light, *Proceedings of SPIE*, 6000, 60,000H1–10 (Boston, MA), 2005.
182. G. A. Ayubi, J. A. Ayubi, J. M. D. Martino, and J. A. Ferrari, Pulse-width modulation in defocused 3-D fringe projection, *Opt. Lett.* 35, 3682–3684, 2010.
183. C. Zuo, Q. Chen, S. Feng, F. Feng, G. Gu, and X. Sui, Optimized pulse width modulation pattern strategy for three-dimensional profilometry with projector defocusing, *Appl. Opt.* 51(19), 4477–4490, 2012.
184. W. Lohry and S. Zhang, 3D shape measurement with 2D area modulated binary patterns, *Opt. Laser Eng.* 50(7), 917–921, 2012.
185. W. Lohry and S. Zhang, Fourier transform profilometry using a binary area modulation technique, *Opt. Eng.* 51(11), 113602, 2012.
186. T. L. Schuchman, Dither signals and their effect on quantization noise, *IEEE Trans. Commun. Technol.* 12(4), 162–165, 1964.

187. W. Purgathofer, R. Tobler, and M. Geiler, Forced random dithering: Improved threshold matrices for ordered dithering, *IEEE International Conference on Image Processing* 2, 1032–1035, (Austin, Texas), November 13–16, 1994.

188. B. Bayer, An optimum method for two-level rendition of continuous-tone pictures, *IEEE International Conference on Communications* 1, 11–15, (Seattle, Washington), June 11–13, 1973.

189. T. D. Kite, B. L. Evans, and A. C. Bovik, Modeling and quality assessment of halftoning by error diffusion, *IEEE International Conference on Image Processing* 9(5), 909–922, (Vancouver, Canada), September 10–13, 2000.

190. R. Floyd and L. Steinberg, An adaptive algorithm for spatial gray scale, *Proceedings of the Society for Information Display*, 17(2), 75–77, 1976.

191. P. Stucki, MECCA multiple-error correcting computation algorithm for bilevel hardcopy reproduction, Tech. Rep., IBM Res. Lab., Zurich, Switzerland, 1981.

192. J. Dai and S. Zhang, Phase-optimized dithering technique for high-quality 3D shape measurement, *Opt. Laser Eng.* 51(6), 790–795, 2013.

193. J. Dai, B. Li, and S. Zhang, Intensity-optimized dithering technique for high-quality 3D shape measurement, *Opt. Laser Eng.* 53, 79–85, 2014.

194. J. Sun, C. Zuo, S. Feng, S. Yu, Y. Zhang, and Q. Chen, Improved intensity-optimized dithering technique for 3D shape measurement, *Opt. Laser. Eng.* 66, 158–164, 2015.

195. Y. Wang, B. Bhattacharya, E. H. Winer, P. Kosmicki, W. H. El-Ratal, and S. Zhang, Digital micromirror transient response influence on superfast 3D shape measurement, *Opt. Laser Eng.* 58, 19–26, 2014.

196. N. Karpinsky and S. Zhang, 3D range geometry video compression with the H.264 codec, *Opt. Laser Eng.* 51(5), 620–625, 2013.

Index

Milton Keynes UK
Ingram Content Group UK Ltd.
UKHW040058071024
449327UK00019B/649